Hot and Cold
Water Supply
Second Edition

Hot and Cold Water Supply
Second Edition

Prepared by

R.H. Garrett, EngTech, MIP RP
*Past President of the National Association of
Plumbing Teachers
Plumber of the Year 1988*

for

British Standards Institution

Blackwell
Science

Blackwell Science Ltd, a Blackwell Publishing company

Editorial offices:
Blackwell Science Ltd, 9600 Garsington Road, Oxford OX4 2DQ, UK
  Tel: +44 (0) 1865 776868
Blackwell Publishing Inc., 350 Main Street, Malden, MA 02148-5020, USA
  Tel: +1 781 388 8250
Blackwell Science Asia Pty, 550 Swanston Street, Carlton, Victoria 3053, Australia
  Tel: +61 (0)3 8359 1011

First published 2000
Reprinted 2001, 2003 (twice), 2004, 2006

ISBN-10: 0-632-04985-5
ISBN-13: 978-0-632-04985-1

Library of Congress Cataloging-in-Publication Data
Garrett, R.H. (Robert H.)
   Hot and cold water supply/prepared by R.H. Garrett.
  —2nd ed.
    p. cm.
   Includes bibliographical references and index.
   ISBN 0-632-04985-5 (pb)
   1. Plumbing. 2. Water-supply engineering. I. Title.
   TH6521 .G37 2000
   696′.1—dc21

A catalogue record for this title is available from the British Library

Set in 10/11.5 Souvenir
by DP Photosetting, Aylesbury, Bucks
Printed and bound in Great Britain
by TJ International Ltd, Padstow, Cornwall

For further information on Blackwell Publishing, visit our website:
www.blackwellpublishing.com

Contents

Acknowledgements

BSI wishes to thank Frank Young for his assistance and the following companies/organizations for their kind permission to use material.

Braithwaite Engineers Ltd
British Coal Corporation
British Gas Plc
Caradon Mira Ltd
Cistermiser Ltd
Conex-Sanbra Ltd
Construction Industry Training Board
Copper Development Association
Corgi
Danfoss Ltd
Department of the Environment, Transport and the Regions
Ecowater Systems Ltd
F W Talbot & Co Ltd
George Fischer Sales Ltd
Gledhill Water Storage Ltd
Glow-worm Ltd
Grundfoss Pumps
Honeywell Control System Ltd
H Warner & Son Ltd
IMI Range Ltd
IMI Santon Ltd
IMI Yorkshire Copper Tube Ltd
IMI Yorkshire Fittings Ltd
Lancashire Fittings
Meter Options Ltd
North West Water Ltd
Polystel Ltd
Polytank Ltd
Reliance Water Controls
The Architectural Press Ltd
The Institute of Plumbing
Water Research Centre
Water Training International
Worcester Heat Systems Ltd

Introduction

The development, construction, installation and maintenance of hot and cold water supply systems are vital areas of concern for public health. Water quality is by turns a political, environmental and technical issue. It is governed by legislation, water regulations, building regulations and technical standards intended to safeguard quality.

This book is a thorough introduction and guide to hot and cold water supply. It is based on a British Standard, BS 6700, which is a specification for design, installation, testing and maintenance of water supply services for domestic buildings. It also includes information on the law and on the water regulations. It is an invaluable work of reference for designers, installers and contractors who work on domestic water supply. It is essential reading for students of plumbing – especially those who are working towards S/NVQ or BTech qualifications. It is also an important aid to technical staff of water undertakers, structural surveyors, architects, building contractors and property services managers.

At present it is not a legal requirement that anyone installing or repairing domestic water services should be properly qualified. However, the installer or repairer can be prosecuted for offences under the water regulations or could be prosecuted under civil law by a householder. With increasing pressure to maintain the highest standards of health and safety and to prevent waste, misuse or contamination of water it is important that technical knowledge is up-to-date.

There is also a strong movement to make water supply practice consistent throughout Europe. In continental Europe hot water systems are supplied from mains pressure rather than from storage. This book helps practitioners to come to terms with these changes.

Hot and Cold Water Supply was written for the British Standards Institution by Bob Garrett, and has benefited from the assistance and comments of other plumbers, designers and teachers. A principle of *Hot and Cold Water Supply* is to support technical information with illustrations and examples. This is especially useful for students and for those who are engaged in training. It is an immensely practical book which makes easy the job of turning theory into action.

This book has been updated to take account of Building Regulations, revisions to the relevant British Standards and the Water Supply (Water Fittings) Regulations 1999.

Chapter 1
General considerations

1.1 Legislation

Water Regulations

In England and Wales, water supply was for many years governed by water byelaws, made by water undertakers under Section 17 of the Water Act 1945, for the prevention of waste, undue consumption, misuse, or contamination of water supplied by the undertaker. Byelaws have now been replaced by the Water Supply (Water Fittings) Regulations 1999 made by the Secretary of State for the Department of the Environment, Transport and the Regions (DETR). These came into force on 1 July 1999.

The new regulations, applicable to England and Wales, are made under Sections 74, 84 and 213(2) of the Water Industry Act 1991 and are similar in content to the previous byelaws, but have been amended to take account of the latest technical advancement and innovation. Unlike previous legislation, these new Water Regulations could only be made after consultation, through the European Commission, with other Member States of the European Community.

In Scotland, water byelaws are made under Section 70 of the Water (Scotland) Act 1980, and in Northern Ireland water regulations have been made under Article 40 of the Water and Sewerage Services (Northern Ireland) Order 1973. Both the byelaws in Scotland and the regulations in Northern Ireland are being revised to come into line with the regulations applicable to England and Wales.

A building owner or occupier can demand a supply of water for domestic purposes provided the relevant requirements of the Water Industry Act 1991 have been complied with and the installation satisfies the requirements of the Water Regulations.

Whilst it is the duty of the water supplier to accede to the owner's demand, the supplier must also uphold the requirements of the Water Regulations and has the right to refuse connection to the mains of any new installation which is not in compliance with them.

To avoid unnecessary dispute with water undertakers, and perhaps lengthy legal proceedings, it is advisable to consult water undertakers about their regulations at an early stage, and particularly their requirements arising from local water or soil characteristics.

Although there is no legal requirement for a person installing or repairing water services to be suitably qualified, anyone who carries out work of this kind can be prosecuted for an offence against current Water Regulations. It should be noted that householders also commit an offence under the regulations if they 'use' a fitting which does not comply with the regulations. Knowledge of the regulations is therefore desirable for both installer and user.

Particular note should be taken of Regulation 5 which requires notice in writing before commencing work on any of the following:

(1) The erection of a building or other structure, not being a pond or swimming pool.
(2) The extension or alteration of a water system on any premises other than a house.
(3) A material change of use of any premises.
(4) The installation of:
     (a) a bath having a capacity, measured to the centre line of overflow, of more than 230 l;
     (b) a bidet with an ascending spray or flexible hose;
     (c) a single shower unit (which may consist of one or more shower heads within a single unit) not being a drench shower installed for reasons of safety or health, connected directly or indirectly to a supply pipe which is of a type specified by the regulator;
     (d) a pump or booster drawing more than 12 l/min, connected directly or indirectly from the supply pipe;
     (e) a unit which incorporates reverse osmosis;
     (f) a water treatment unit which produces a waste discharge or which requires the use of water for regeneration or cleaning;
     (g) a reduced pressure zone valve assembly or other mechanical device for protection against a fluid which is in fluid-risk category 4 or 5;
     (h) a garden watering system unless designed to be used by hand; or
     (i) any water system laid outside a building and either less than 750 mm or more than 1350 mm below ground level.
(5) The construction of a pond or swimming pool with a capacity greater than 10 000 l which is designed to be replenished by automatic means and is to be filled with water supplied by a water undertaker.

Under Regulation 7 any person who on summary conviction is found guilty of Water Regulations contraventions is liable to a fine, for each separate offence, not exceeding level three on the standard scale. At the time of writing this means a fine of up to £1000 for each offence. In Scotland the fine is level four on the standard scale.

Water undertakers are encouraged under Water Regulations to set up approved contractors' schemes within their area of supply. The scheme, it is suggested, will be similar to that already being operated by Anglia Water and will require that approved contractors certify to the undertaker that water fittings installed are in compliance with the regulations. Approved contractors will be excused from some of the notifications listed above, but by no means all of them.

The scheme will be of benefit to consumers who, in any proceedings against them, can show 'that the work was carried out by, or under the direction of, an approved contractor, and that the contractor certified to them that the water fittings complied with the requirements of the Regulations'.

New measures for water conservation are encouraged by the reintroduction of dual-flushing cisterns, and the new provisions will permit non-siphonic flushing methods for WCs. As from 1 January 2001, maximum WC flushing volumes will be reduced from 7.5 l to 6 l.

A Secretary of State Specification for WC Suite Performance has been developed and will require extensive testing for all WCs whether they be siphonic or non-siphonic. Previous byelaw requirements will remain in place until the new specification comes into effect on 1 January 2001.

Pressure flushing valves will be permitted (but not in a house) for use in the flushing of both WCs and urinals which may be connected directly to either a supply pipe or a distributing pipe, but they must have backflow protection fitted.

Backflow protection requirements have been revised. The new regulations introduce, in Schedule 1, five fluid-risk categories (rather than the three under previous byelaws) to bring us into line with European practices. At the same time the backflow prevention devices are categorized differently in a new Secretary of State specification.

We have been used to safety devices for unvented hot water heaters for a long time. Thermostatic control and temperature relief valves have, of course, been required by the Building Regulations for a good many years. Water Regulations now duplicate this requirement.

In Schedule 2 to Regulation 4(3), Paragraph 18 it says 'Appropriate vent pipes, temperature control devices and combined temperature pressure relief valves shall be provided to prevent the temperature of the water within a secondary hot water system from exceeding 100°C'.

Water systems are required under Paragraph 13 of Schedule 2 to be tested, flushed and, where necessary, disinfected before use. This is not new, but more emphasis is given to the need for these important aspects of water installations.

It should be added that Water Regulations are *not* made for the specific protection of people or property, but solely for the prevention of waste, undue consumption, misuse and contamination of water supplied by water undertakers:

'waste'   means water which flows away unused, e.g. from a dripping tap or hosepipe left running when not in use, or from a leaking pipe.

'undue consumption'   means water used in excess of what is needed, e.g. full bore tap to wash hands when half or quarter flow will suffice, or automatic flushing cistern that flows even when urinals are not in use (at night).

'misuse'   means water used for purposes other than that for which it is supplied, e.g. use of garden sprinkler when paying only for domestic use or taking supply from domestic premises for industrial or agricultural use.

'contamination'   means pollution of water by any means, e.g. by cross connection between public and private supplies or by backflow through backsiphonage.

'erroneous measurement'   means incorrect meter reading, e.g. connections made which may not be detected by the meter.

Guidance on particular Water Regulation matters may be sought from local water undertakers. Their inspectors are trained in the application of Water Regulations.

The Water Regulations Advisory Scheme (WRAS) has published a guide to the application and interpretation of Water Regulations. It contains useful information and background knowledge for those concerned with water services.

1.2   Guidance and approval of water fittings

The Water Research Centre, through its Water Regulations Advisory Scheme (WRAS), also operates a voluntary national scheme for the testing and approval of water fittings. Fittings which pass the Centre's tests are listed in its publication *Water Fittings and Materials Directory*, together with the names and addresses of manufacturers and any applicable installation requirements. Thus, the connection and use of any listed fitting carries with it virtual certainty of acceptance by water undertakers.

Building Regulations

Building Regulations in England and Wales are made under the Building Act 1984 to cover the health and safety of people in and around buildings. In general the Building Regulations do not include requirements for water, gas and electricity, but there are some specific water-related provisions. These are:

- The provision of wash basins in the vicinity of water closets (WCs), hot and cold water to wash basins and provision for cleaning of wash basins and WCs (Approved Document G1).
- The provision of baths and showers and the supply of hot and cold water to them (Approved Document G2).
- Provision for precautions for the safety of hot water systems and in particular prevention of explosion (Approved Document G3).
- Requirements for the control of energy in heating and hot water systems and the insulation of hot water pipes and storage vessels.

The responsibility for control of building work under the Building Regulations lies with local authorities and their building control officers (building inspectors) or other approved inspectors such as the National House Building Council (NHBC).

Scotland and Northern Ireland have similar building legislation, namely the Building Standards (Scotland) Regulations and the Building (Northern Ireland) Regulations.

The Health and Safety at Work etc. Act 1974

This Act provides for securing the health, safety and welfare of persons at work, for controlling the use and storage of dangerous substances and for the control of certain emissions into the atmosphere. The Health and Safety at Work etc. Act 1974 is an enabling act under which many safety regulations are made. Some of those relevant to this book include:

- The Workplace (Health, Welfare and Safety) Regulations which regulated the provision of drinking water and sanitary accommodation in the workplace.
- The Gas Safety (Installation and Use) Regulations, used to control the installation of gas appliances, provide for the registration of gas installers and require that all gas-fitting operatives shall be competent in the area of gas work that they do.

1.3   Scope of the standard

BS 6700 specifies requirements and gives recommendations for the design, installation, alteration, commissioning and testing, and maintenance of hot water supply, cold water supply and frost precautions.

The following are not included although, in parts, the standard may apply to them: hot water systems whose temperature exceeds 100°C; central heating systems; fire fighting; and water for industrial purposes.

1.4   Definitions

From BS 6700

The following definitions are used:

backflow   a flow of water in the opposite direction to that intended. It includes backsiphonage, which is backflow caused by siphonage.

building   any structure (including a floating structure) whether of a permanent character or not, and whether movable or immovable, connected to the water supplier's mains.

cavity wall   any wall whether structural or partition that is formed by two upright parts of similar or dissimilar building materials suitably tied together with a gap formed between them which may be (but need not be) filled with insulating material.

chase   a recess that is cut into an existing structure.

cover   a panel or sheet of rigid material fixed over a chase, duct or access point, of sufficient strength to withstand surface loadings appropriate to its position.
NOTE   Except where providing access to joints or changes of direction (i.e. at an inspection access point) a cover may be plastered or screeded over.

duct   an enclosure designed to accommodate water pipes and fittings and other services, if required, and constructed so that access to the interior can be obtained either throughout its length or at specified points by removal of a cover or covers.

dwelling premises, buildings or part of a building providing accommodation, including a terraced house, a semi-detached house, a detached house, a flat in a block of flats, a unit in a block of maisonettes, a bungalow, a flat within any non-domestic premises, a maisonette in a block of flats, or any other habitable building and any caravan, vessel, boat or houseboat connected to the water supplier's mains.

inspection access point a position of access to a duct or chase whereby the pipe or pipes therein can be inspected by removing a cover which is fixed by removable fastenings but does not necessitate the removal of surface plaster, screed or continuous surface decoration.

removable fastenings fastenings that can be removed readily and replaced without causing damage including turn buckles, clips, magnetic or touch latches, coin-operated screws and conventional screws, but do not include nails, pins or adhesives.

sleeve an enclosure of tubular or other section of suitable material designed to provide a space through an obstruction to accommodate a single water pipe and to which access to the interior can be obtained only from either end of such sleeve.

tap size designations numbers directly related to the nominal size of the thread on the inlet of the tap, which in turn is unchanged from the nominal size in inches before metrication, e.g. $\frac{1}{2}$ nominal size tap means a tap with an inlet having a G $\frac{1}{2}$ thread.

walkway or crawlway an enclosure similar to a duct, but of such size as to provide access to the interior by persons through doors or manholes and which will accommodate water pipes and fittings and other services if required.

From Water Regulations

A further list of definitions relating to Schedule 2 of the Water Supply (Water Fittings) Regulations 1999:

backflow means flow upstream, that is in a direction contrary to the intended normal direction of flow, within or from a water fitting.

cistern means a fixed container for holding water at atmospheric pressure.

combined feed and expansion cistern means a cistern for supplying cold water to a hot water system without a separate expansion cistern.

combined temperature and pressure relief valve means a valve capable of performing the function of both a temperature relief valve and a pressure relief valve.

concealed water fitting means a water fitting that:

(a)   is installed below ground;
(b)   passes through or under any wall, footing or foundation;
(c)   is embedded in any wall or solid floor;
(d)   is enclosed in any chase or duct; or
(e)   is in any other position which is inaccessible or renders access difficult.

contamination   includes any reduction in chemical or biological quality of
water due to raising its temperature or the introduction of polluting substances.

distributing pipe   means any pipe (other than a warning, overflow or
flush pipe) conveying water from a storage cistern, or from hot water
apparatus supplied from a cistern and under pressure from that cistern.

expansion cistern   means a cistern connected to a water heating system
which accommodates the increase in volume of that water in the system
when the water is heated from cold.

expansion valve   means a pressure-activated valve designed to release
expansion water from and unvented water heating system.

overflow pipe   means a pipe from a cistern in which water flows only
when the water level in the cistern exceeds its normal maximum level.

pressure flushing cistern   means a WC flushing device that utilizes the
pressure of water within the cistern supply pipe to compress the air and
thus increase the pressure of water available for flushing a WC pan.

pressure relief valve   means a pressure-activated valve which opens
automatically at a specified pressure to discharge fluid.

primary circuit   means an assembly of water fittings in which water circulates between a boiler or other source of heat and a primary heat
exchanger inside a hot water storage vessel.

secondary circuit   means an assembly of water fittings in which water
circulates in supply pipes or distributing pipes to and from a hot water
storage vessel.

secondary system   means that part of any hot water system compris-
ing the cold feed pipe, any hot water storage vessel, water heater and
flow and return pipework from which hot water is conveyed to all points
of draw-off.

servicing valve   means a valve for shutting off the flow of water in a pipe
connected to a water fitting for the purpose of maintenance or service.

spill-over level   means the level at which the water in a cistern or sanitary
appliance will first spill over if the inflow of water exceeds the outflow
through any outflow pipe and any overflow pipe.

stopvalve   means a valve, other than a servicing valve, for shutting off the
flow of water in a pipe.

supply pipe   means so much of any pipe as is not vested in the water
undertaker.

temperature relief valve   means a valve which opens automatically at a specified temperature to discharge water.

terminal fitting   means a water discharge point.

unvented hot water storage vessel   means a hot water storage vessel that is not provided with a vent pipe but is fitted with safety devices to control primary flow, prevent backflow, control working pressure and accommodate expansion.

vent pipe   means a pipe open to the atmosphere which exposes the system to atmospheric pressure at its boundary.

warning pipe   means an overflow pipe whose outlet is located in a position where the discharge of water can be readily seen.

Graphical symbols

See figure 1.1.

Symbols used in this book are, where possible, based on those given in BS 1192: Part 3. However, there are many components not included in BS 1192, so symbols from other sources have been used.

Symbol	Description	BS 1192 Ref.	Application
▶	draw-off tap (valve port)	7.3 7.4	
	shower head		
△	sprinkler head (spray outlet)	7.212	
	float-operated valve (balloolt)	7.309	
	float switch (hydraulic type)		
	float switch (magnetic type)		
	filter or screen		
⋈	supply stopvalve (SV)	7.301	
⋈	servicing valve (SV)		
water	water meter	7.226	

Figure 1.1   Graphical symbols and abbreviations

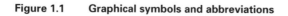

continued

Symbol	Description	BS 1192 Ref.	Application
	draining valve (BS 1192) (drain valve) (drain cock)	7.305	
	hose connection (used in this book as a draining valve)	7.3E8	
	line strainer	7.218	
	pressure reducing valve (small end denotes high pressure)	7.307	
	expansion vessel	7.610	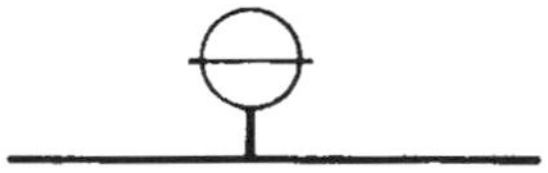
	pressure relief valve (expansion relief valve)	7.316	
	check valve or non-return valve (NRV)	7.306	 nrv
	double check valve assembly		
	combined check and anti-vacuum valve (check valve and vacuum breaker)		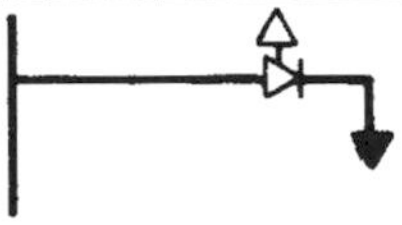
	air inlet valve		

Figure 1.1   Graphical symbols and abbreviations continued

continued

Symbol	Description	BS 1192 Ref.	Application
 CWSC	cold water storage cistern (storage and feed cistern) (feed cistern)		
 F&ExC	feed and expansion cistern		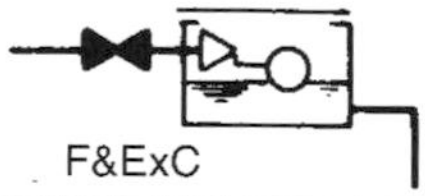
 HWC	hot water storage cylinder or tank (plan)		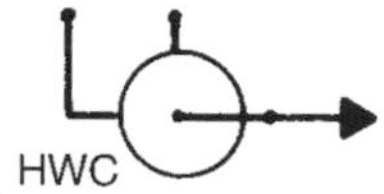
 HWC	direct hot water storage cylinder or hot store vessel (elevation)		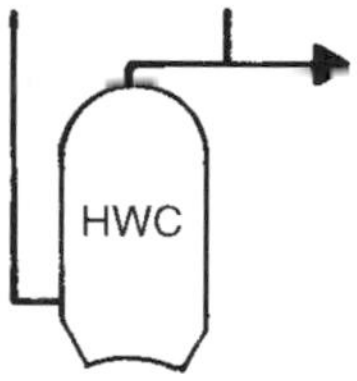
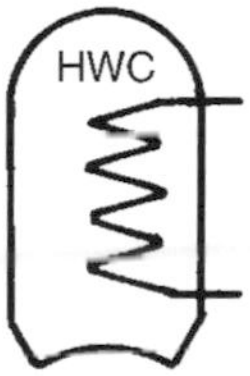 HWC	Indirect hot water storage cylinder or hot store vessel (elevation)		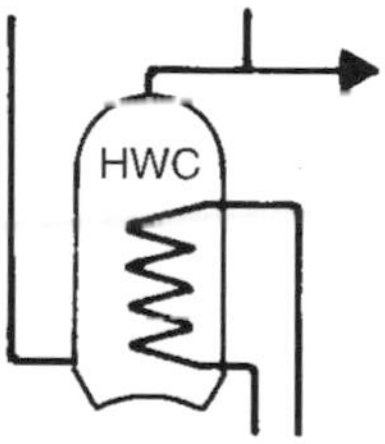
	boiler (elevation)		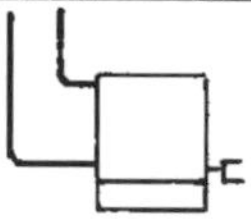
	temperature relief valve		
 Y	tundish		

Figure 1.1   Graphical symbols and abbreviations continued

continued

Symbol	Description	BS 1192 Ref.	Application
	pump (centrifugal)	7.609	
	pump (circulating)	7.608	
	automatic air vent	7.224	
	insulation	2.209 2.210	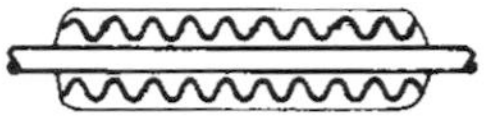
 s	sink (elevation)		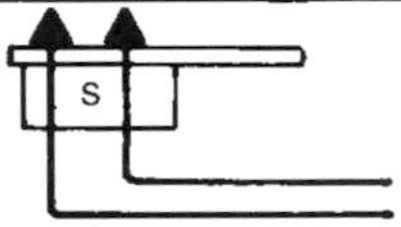
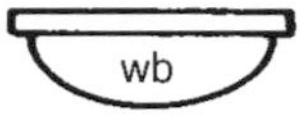 wb	wash basin (elevation)		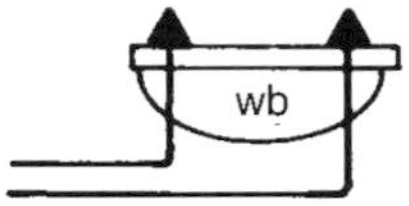
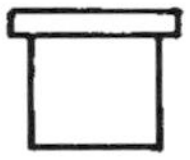	bidet (elevation)		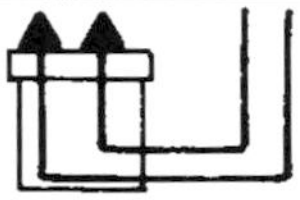
	bath (elevation)		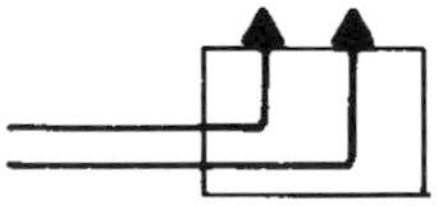
 wc	water closet (WC) (elevation)		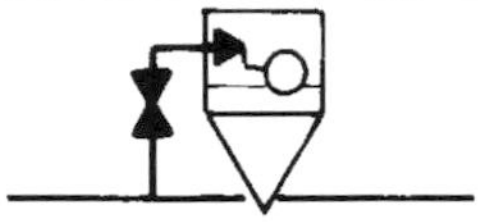
	urinal bowl (elevation)		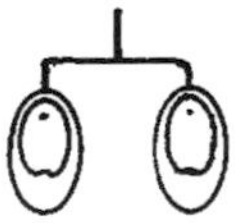

Figure 1.1 **Graphical symbols and abbreviations** **continued**

1.5 Materials

Selection and use of materials is dealt with in greater depth in chapter 11.

Choice of material for a particular water installation may be determined by the following factors:

- effect on water quality;
- cost, service life and maintenance needs;
- for metallic pipes internal and external corrosion (particularly from certain waters);
- compatibility of materials;
- ageing, fatigue and temperature effects, especially in plastics;
- mechanical properties and durability;
- vibration, stress or settlement;
- internal water pressures.

The water supplier should be consulted at an early stage, particularly about the choice of materials in relation to the character of the water supply and ground conditions. Some waters are aggressive to certain pipes as are certain types of soil. The water supplier should be able to advise on local conditions, and give guidance on the suitability and application of proposed materials.

Pipes and fittings should be used only within the limits stated in relevant British Standards and in accordance with any manufacturers' recommendations, and the requirements of Water Regulations should be met. Installations should be capable of operating effectively under the conditions they will experience in service.

Pipes, joints and fittings should be capable of withstanding sustained temperatures as shown in table 1.1, without damage or deterioration.

Table 1.1   Temperature limits for pipes and fittings

Cold water installations	40°C
Hot water installations	95°C (with occasional short-term excursions up to 100°C)
Discharge pipes connected to temperature or expansion relief valves in unvented hot water systems	125°C

Pipes, joints and fittings of dissimilar metals should not be connected together unless precautions are taken to prevent corrosion. This is particularly important in below ground installations where conditions are often conducive to corrosion.

Table 1.2 gives a selection of pipes of various materials along with a range of comparative internal and external diameters.

Detailed information on pipes and fittings can be see in chapter 11.

Table 1.2   Equivalent pipe sizes

Equivalent pipe sizes | | | | | **Rigid pipes** | | | | | | | | |
|---|---|---|---|---|---|---|---|---|---|---|---|---|---|---|
| Comparative internal size in millimetres | Copper* to BS EN 1057 | | Stainless Steel to BS 4127: Part 2 | | Steel (screwed) (galvanized or black)* to BS 1387 (medium grade) | | | Grey iron to BS 4622 | | Ductile iron to BS 4772 | | Asbestos cement* to BS 486 (Class 25) | | Comparative internal size in inches |
| | ID mm | BS nominal size OD mm | ID mm | BS nominal size OD mm | BS nominal size ID mm | OD mm | Thread designation BS 21 inches | BS nominal diameter ID mm | OD mm | BS nominal diameter ID mm | OD mm | BS nominal diameter ID mm | OD mm | |
| 4.5 | 4.8 | 6 | 4.8 | 6 | — | — | — | — | — | — | — | — | — | $\frac{1}{8}$ |
| 6 | 6.8 | 8 | 6.8 | 8 | — | — | — | — | — | — | — | — | — | $\frac{1}{4}$ |
| 8 | 8.8 | 10 | 8.8 | 10 | 8 | 13.6 | $\frac{1}{4}$ | — | — | — | — | — | — | $\frac{1}{4}$ |
| 10 | 10.8 | 12 | 10.8 | 12 | 10 | 17.1 | $\frac{3}{8}$ | — | — | — | — | — | — | $\frac{1}{4}$ |
| 13 | 13.6 | 15 | 13.6 | 15 | 15 | 21.4 | $\frac{1}{2}$ | — | — | — | — | — | — | $\frac{3}{8}$ |
| 15 | 16.4 | 18 | 16.6 | 18 | — | — | — | — | — | — | — | — | — | $\frac{1}{2}$ |
| 20 | 20.2 | 22 | 20.6 | 22 | 20 | 26.9 | $\frac{3}{4}$ | — | — | — | — | — | — | $\frac{3}{4}$ |
| 25 | 26.2 | 28 | 26.4 | 28 | 25 | 33.8 | 1 | — | — | — | — | — | — | 1 |
| 32 | 32.6 | 35 | 33 | 35 | 32 | 42.5 | $1\frac{1}{4}$ | — | — | — | — | — | — | $1\frac{1}{4}$ |
| 40 | 39.6 | 42 | 39.8 | 42 | 40 | 48.4 | $1\frac{1}{2}$ | — | — | — | — | — | — | $1\frac{1}{2}$ |
| 50 | 51.8 | 54 | — | — | 50 | 59.3 | 2 | — | — | — | — | 50 | 69 | 2 |
| 63 | 64.6 | 67 | — | — | 65 | 80.1 | $2\frac{1}{2}$ | — | — | — | — | — | — | $2\frac{1}{2}$ |
| 75 | 73.1 | 76.1 | — | — | 80 | 88.8 | 3 | 80 | 98 | 80 | 98 | 75 | 96 | 3 |
| 100 | 105 | 108 | — | — | 100 | 113.9 | 4 | 100 | 118 | 100 | 118 | 100 | 122 | 4 |
| 125 | 130 | 133 | — | — | 125 | 139.6 | 5 | — | — | — | — | 125 | — | 5 |
| 150 | 155 | 159 | — | — | 150 | 165.1 | 6 | 150 | 170 | 150 | 170 | 150 | 177 | 6 |
| 200 | – | – | — | — | — | — | — | 200 | 222 | 200 | 222 | 200 | 240 | 8 |
| 250 | – | – | — | — | — | — | — | 250 | 274 | 250 | 274 | 250 | 295 | 10 |
| 300 | – | – | — | — | — | — | — | 300 | 326 † | 300 | 326 † | 300 | 356 † | 12 |

Note   Some intermediate sizes have been omitted.
* In some materials only one grade is shown.
† Larger sizes than 300 mm have been excluded.

Table 1.2 continued

Flexible pipes

Comparative internal size in millimetres	Unplasticized (PVC-U) to BS 3505 (Class E)			Propylene copolymer to BS 4991		Polyethylene to BS 6572 medium density BLUE		to BS 6730 medium density BLACK		to BS 1972 low density Class C		Comparative internal size in inches
	BS nominal size ID inches	OD mm	ID mm	BS nominal size ID inches	OD mm	ID mm	BS nominal size OD mm	ID mm	BS nominal size OD mm	OD mm	BS nominal size ID inches	
4.5	—	—	—	—	—	—	—	—	—	—	—	1/8
6	—	—	—	—	—	—	—	—	—	—	—	1/4
8	—	—	—	1/4	13.6	—	—	—	—	—	—	1/4
10	—	—	—	3/8	17.1	—	—	—	—	17.1	3/8	3/8
13	3/8	17.1	15.3	1/2	21.3	—	—	—	—	21.3	1/2	1/2
15	1/2	21.3	17.5	—	—	15.1	20	15.1	20	—	—	—
20	3/4	26.7	21.7	3/4	26.7	20.4	25	20.4	25	26.7	3/4	3/4
25	1	33.5	28.3	1	33.5	25	32	26	32	33.5	1	1
32	1¼	42.2	34.8	1¼	42.2	—	—	—	—	42.3	1¼	1¼
40	1½	48.2	41.0	1½	48.3	40.8	50	40.8	50	48.3	1½	1½
50	2	60.3	51.3	2	60.3	51.4	63	51.4	63	60.3	2	2
63	—	—	—	2½	75.3	—	—	—	—	—	—	2½
75	3	88.9	76.5	3	88.9	—	—	—	—	—	—	3
100	4	114.3	97.7	4	114.3	—	—	—	—	—	—	4
125	5	140.2	120	—	—	—	—	—	—	—	—	5
150	6	168.2	144	6	168.3	—	—	—	—	—	—	6
200	8	219.1	190.9	8	219.5	—	—	—	—	—	—	8
250	10	273	138	10	273.4	—	—	—	—	—	—	10
300	12	323.8†	282.2†	12†	323.8†	—	—	—	—	—	—	12

continued

Note Some intermediate sizes have been omitted.
* In some materials only one grade is shown.
† Larger sizes than 300 mm have been excluded.

Table 1.2 continued

Flexible pipes continued

Comparative internal size in millimetres	Polybutylene (PB) to BS 7291: Parts 1 and 2				Cross-linked polyethylene (PE-X) to BS 7291: Parts 1 and 3				Chlorinated polyvinyl chloride (PVC-C) to BS 7291: Parts 1 and 4		Comparative internal size in inches
	Consistent with BS 5556		Consistent with BS 2871		Consistent with BS 5556		Consistent with BS 2871		BS nominal size		
	ID mm	OD BS nominal size mm	ID mm	OD BS nominal size mm	ID mm	OD BS nominal size mm	ID mm	OD BS nominal size mm	ID mm	OD mm	
6	6.7	10	6.7	10	6.7	10	6.7	10	—	—	$\frac{1}{4}$
8	8.7	12	8.7	12	8.7	12	8.7	12	8.5	12	$\frac{3}{8}$
12	12.7	16	12.2	15	12.7	16	12.2	15	12	16	$\frac{1}{2}$
15	15.9	20	14.2	18	15.9	20	14.2	18	15.8	10	
20	20.1	25	17.7	22	20.1	25	17.7	22	19.9	25	$\frac{3}{4}$
25	26.1	32	22.5	28	26.1	32	22.5	28	25.6	32	1
32	—	—	28.3	35	—	—	28.3	35	32.2	40	$1\frac{1}{4}$
40	—	—	—	—	—	—	—	—	40.3	50	$1\frac{1}{2}$
50	—	—	—	—	—	—	—	—	50.85	63	2

Note PB pipes and PE-X pipes are also available in a range of sizes consistent with those specified in BS 2871. These are 10, 12, 15, 18, 22, 28 and 35 mm.

1.6 Initial procedures

Hot and cold water installations should be designed to conform with Water Regulations (byelaws in Scotland) and with Building Regulations, with consideration given to economic maintenance of the installation throughout its working life. Particular attention should be given to the prevention of bacterial contamination in both hot and cold water services. Bacterial contamination is more likely to occur in buildings of multiple occupation. Temperatures of both hot and cold water should be controlled to avoid conditions which encourage bacterial growth. Guidance on the preservation of water quality is given in chapter 6.

Attention should also be given to the insulation and control of temperatures for the conservation of energy.

Preliminary design factors are as follows:

- flow requirements and estimated likely consumption;
- location of available water supply;
- quality, quantity and pressure of available supply;
- quantity of storage required;
- requirements of Building Regulations and Water Regulations;
- ground conditions, e.g. subsidence/contamination of site;
- liaison with other parties;
- the specific requirements of the water supplier.

The design should include provision for appliances likely to be added later.

Pipe sizes

Since about 1970 pipe sizes have gradually been changing from imperial to metric measurements. Pipes in some materials were metricated quickly, e.g. copper, while others even now retain their imperial identification.

With metrication came a move away from designating pipe sizes by the inside diameter towards the use of the outside diameter.

Currently European Standards are being developed and these will gradually be seen to replace many of our existing British Standard specifications.

Table 1.2 provides a comparison between the sizes of pipes of different materials, using inside diameters as a base, because it is the inside diameter which determines the water-carrying capacity of the pipe.

Polyethylene tube to BS 1972 is still available. Its use is extensive in farming and agriculture. It was extensively used for supply pipes to dwellings until the advent of medium-density tube to BS 6572.

Polyethylene tube to BS 3284 is no longer available and the standard is withdrawn. However, large quantities of this tube are still in use.

Polyethylene tube is commonly used in sizes above 63 mm. There is currently no British Standard but some larger sized tubes have been approved for use under the WRAS Water Fittings Testing Scheme.

Availability of water supplies

Water supplies are available from:

(1) Nearby public water main at cost to owner or within contract price. If not readily available, the water undertaker must provide mains at the expense of the owner or applicant. It is important that the water supplier is consulted at an early stage.
Note Mains may not be laid until road line and kerb level are permanently established.

(2) Suitable and available supply pipe (not favoured by water suppliers who generally insist that each premises has a separate supply pipe).

(3) Private source. Consider the condition and purity of the water. Chemical and bacterial analyses are advisable. Approval is needed from the local public health authority for drinking water supplies, and a licence to abstract may be required from the water authority.
Note 1 If private supply and public supply are taken to a single property, the water undertaker must be informed and regulations complied with. Water from a private source must not be connected to a supply pipe from a water supplier's main.
Note 2 The water supplier may require the supply to be metered.

Water suppliers will need full details of non-domestic water supply requirements to assess the likely demand and effects on water mains and other users in the locality (see figure 1.2).

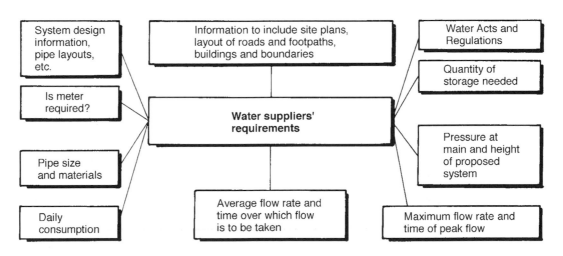

Figure 1.2   **Water suppliers' requirements**

Ground conditions to be considered

(1) Likelihood of contamination of site (local authorities may provide information).

(2) Likelihood of subsidence and other soil movement from:

(a) mining;
(b) vibration from traffic (consider increased depth);
(c) moisture swelling and contraction;
(d) building settlement.

Laying pipes outside buildings

When laying underground service pipes make allowances for pipe materials, jointing and methods of laying. Consider methods of passing pipes through walls. Pipes must be free to deflect.

As far as possible underground pipes:

- should be laid at right angles to the main;
- should be laid in straight lines to facilitate location for repairs, but with slight deviation to adjust to minor ground movement;
- should not be laid under surfaced footpaths or drives;
- should be laid at a minimum depth of 750 mm to avoid frost and other damage, and not deeper than 1350 mm to permit reasonable access.

The use of external pipes above ground should be avoided. Where unavoidable, pipes should be lagged with water-proofed insulation and provision made to drain pipes as a precaution against frost.

Liaison and consultation

From an early stage in the design process, the designer should consult with others involved in the design, installation and use of the system (see figure 1.3).

Where work is to be carried out in a public highway, the highway authority and all private/public utility undertakers should be notified and all

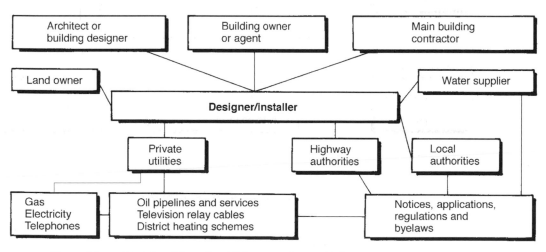

Figure 1.3   **Designers'/installers' liaisons and consultations**

relevant notices completed and lodged with those likely to be affected. These notices should include any drawings and other details of work to be done.

The designer or installer should provide full working drawings, including precise location of all pipe runs, method of ducting, description of all appliances, valves and other fittings, methods of fixing, protections and precautions.

Water pipes of metal should be arranged to permit equipotential bonding to the main electrical earth.

The programme of work should consider:

- method of construction;
- sequence of events including handover to owner;
- coordination of services;
- time needed for construction and services works;
- size and position of incoming services.

Water services to buildings should be coordinated with other services and laid in an orderly sequence and at a line and level that will readily permit maintenance at a later date. On new sites the recommendations of the *National Utilities Group Publication No 6* should be followed.

Chapter 2
Cold water supply

2.1 Drinking water

Under the Water Supply (Water Fittings) Regulations 1999 every dwelling is required to have a wholesome water supply, and this should be provided in sufficient quantities for the needs of the user, and at a temperature below 20°C. The most important place to provide drinking water in dwellings is at the kitchen sink (see figure 2.1). However, because there is a likelihood that all taps in dwellings will be used for drinking, they should all be connected in such a way that the water remains in potable condition. This means that all draw-off taps in dwellings should either be connected direct from the mains supply, or from a storage cistern that is 'protected'.

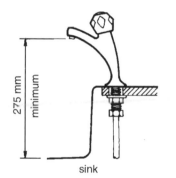

Positioned so that outlet is at least 275 mm above bottom of bowl, to allow buckets and other utensils to be filled.

Outlets designed to make hosepipe connections difficult.

Figure 2.1   **Tap at kitchen sink**

Drinking water supplies should also be provided in suitable and convenient locations in places of work such as offices and commercial buildings, particularly where food and drink are prepared or eaten. If no such locations exist, drinking water should be provided near but not in toilets. However, drinking water fountains may be installed in toilet areas, provided they are sited well away from WCs and urinals and comply with the requirements of BS 6465: Part 1.

Where drinking water taps are fitted in places of work along with taps for other purposes, all taps should be labelled 'drinking water' or 'non-drinking water' as appropriate.

To avoid stagnation drinking water points should not be installed at the ends of long pipes where only small volumes of water are likely to be drawn off, but rather be fitted where there is a high demand downstream.

As far as possible pipe runs to drinking water taps should not follow the routes of space heating or hot water pipes or pass through heated areas. Where this is unavoidable, both hot and cold pipes should be insulated.

In tall buildings where drinking water is needed above the pressure limit of the mains water supply, it may be necessary for the water to be pumped to the higher level. In such cases any drinking water must be supplied from a 'protected' drinking water cistern or from a drinking water header (see later in figure 2.35).

Where a water softener of the base exchange type is installed, connections to drinking water taps should be made upstream of the softener so they do not receive softened water. It should also be noted that softening water increases the potential of the water to dissolve metals. Of particular concern is plumbo solvency where there is a lead pipe that will be reactive to softened water.

2.2   Cold water systems

Systems in dwellings

Water may be supplied to cold taps either directly from the mains via the supply pipe or indirectly from a protected cold water storage cistern. In some cases a combination of both methods of supply may be the best arrangement.

Factors to consider when designing a cold water system should take into account the available pressure and reliability of supply, particularly where any draw-off point is at the extreme end of a supply pipe or situated near the limit of mains pressure.

Figures 2.2. and 2.3. give characteristics of 'direct' and 'indirect' cold water systems and should be useful when considering a system for a particular application.

Cistern omitted where hot water is supplied from: (a) an unvented hot water system; or (b) mains-fed instantaneous water heater.

High pressure supply is more suitable for instantaneous type shower heaters, hose taps and mixer fittings used in conjunction with a high pressure (unvented) hot water supply.

Expensive dual-flow mixer fittings required if used in conjuction with a low pressure (vented) hot water supply.

All taps are supplied under mains pressure and are therefore suitable for drinking and food preparation.

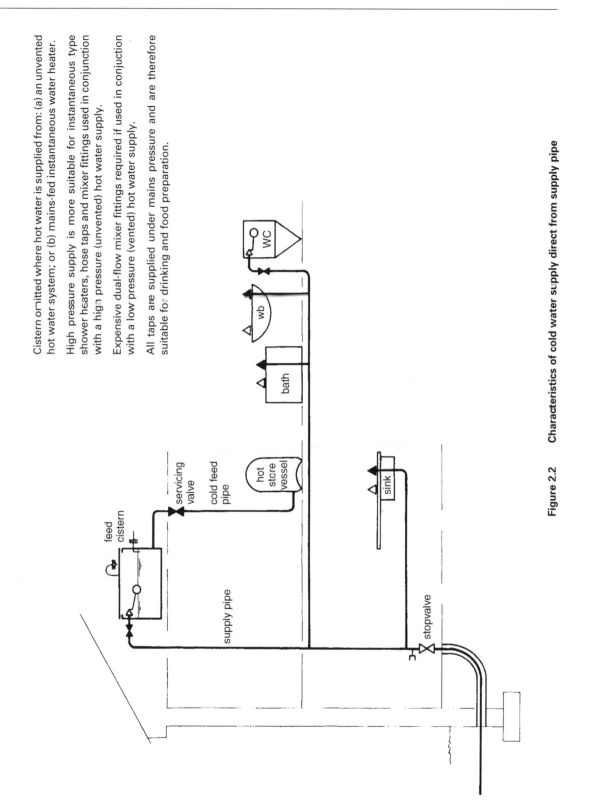

Figure 2.2   Characteristics of cold water supply direct from supply pipe

Risk of frost damage in roof space.

Reserve supply of water available in case of mains failure.

Cistern provides additional protection against contamination of mains.

Pressure available from storage may not be sufficient for some types of tap or shower, e.g. power shower.

Cistern must be continuously protected against the entry of any contaminant; cistern may need to be replaced occasionally.

Space occupied and cost of cistern, structural support and additional pipework must be considered.

Constant low supply pressure reduces the risk and rate of leakage and is suitable for supply to mixer fittings for low pressure (vented) hot water supply.

At least one tap supplied directly from supply pipe for drinking and cooking purposes.

Reduced risk of water hammer and noise from outlets, but additional noise may be generated by the float-operated valve controlling the supply into the cistern.

Drinking water quality at kitchen sink without the need to pass through storage cistern.

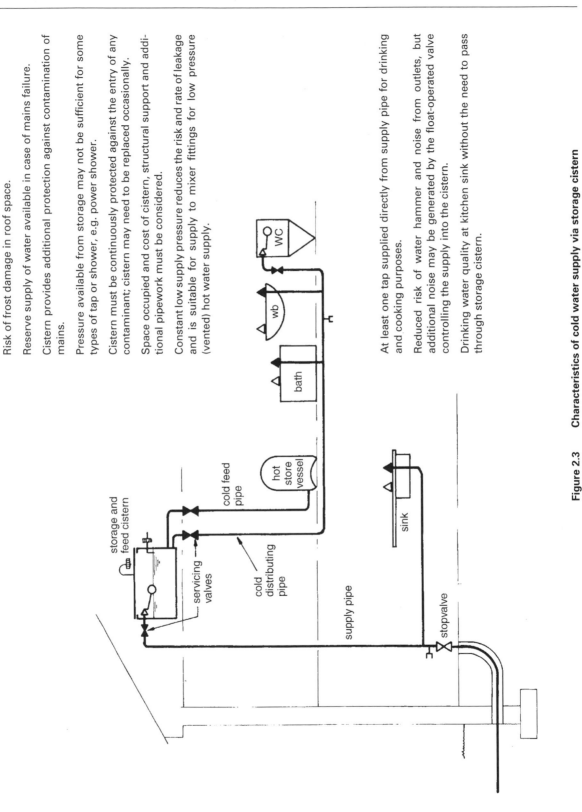

Figure 2.3   Characteristics of cold water supply via storage cistern

Systems in buildings other than dwellings

In the case of small buildings where the water consumption is likely to be similar to that of a dwelling, the characteristics in figures 2.2 and 2.3 should be considered.

For larger buildings such as office blocks, hostels and factories it will usually be preferable for all water, except drinking water, to be supplied indirectly from a cold water storage cistern or cisterns. Drinking water should be taken directly from the water supplier's main wherever practicable (see figure 2.4).

Preferred method of supply for drinking water.

The method of distribution is related to the size and use of the building, and the number and positions of appliances served.

Secondary backflow protection recommended at all floor levels.

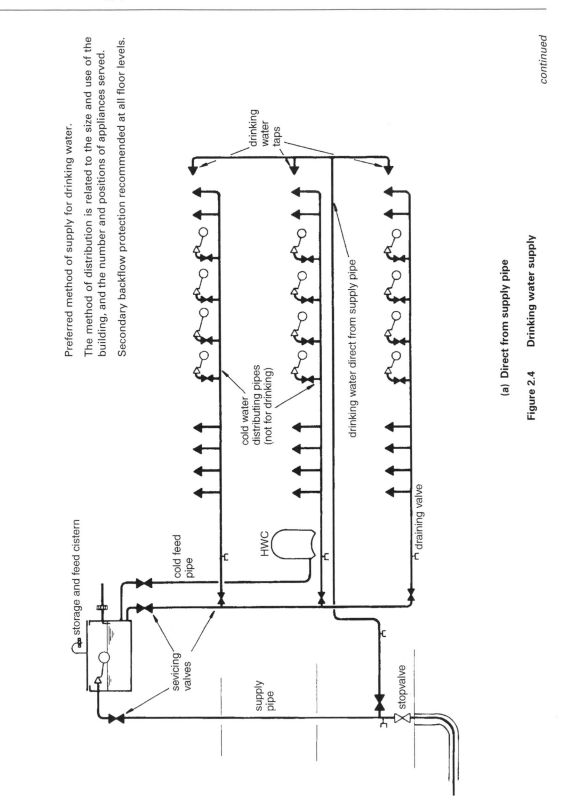

(a) **Direct from supply pipe**

Figure 2.4   **Drinking water supply**

continued

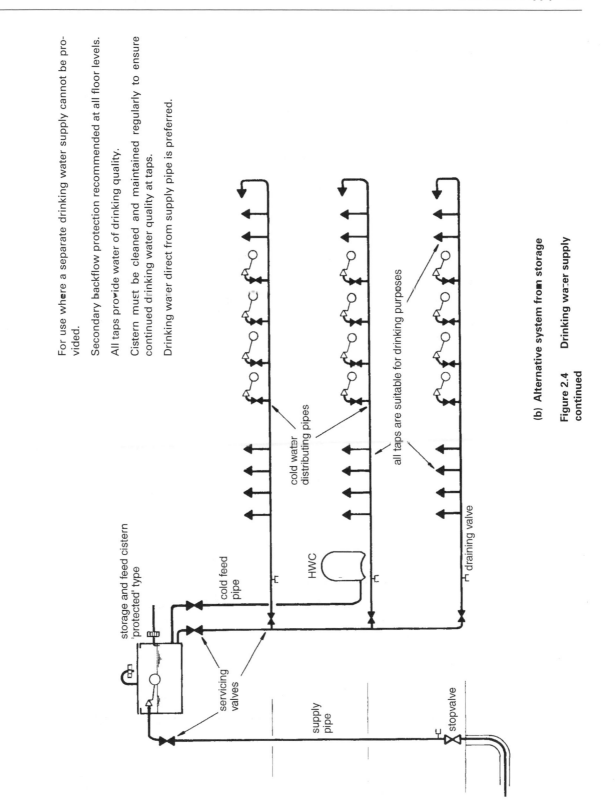

For use where a separate drinking water supply cannot be provided.

Secondary backflow protection recommended at all floor levels.

All taps provide water of drinking quality.

Cistern must be cleaned and maintained regularly to ensure continued drinking water quality at taps.

Drinking water direct from supply pipe is preferred.

storage and feed cistern 'protected' type

cold feed pipe

HWC

servicing valves

supply pipe

stopvalve

draining valve

cold water distributing pipes

all taps are suitable for drinking purposes

(b) Alternative system from storage

Figure 2.4   Drinking water supply
continued

Pumped systems

Where the height of the building lies above statutory levels or when the available pressure is insufficient to supply the whole of a building and the water supplier is unable to increase the supply pressure in the supplier's mains, consideration should be given to the provision of a pumped cold water supply cistern.

It is important that the water supplier is consulted and written consent received before fitting any pump. The water supplier will wish to ensure that Water Regulations are complied with in respect of backflow risk, and that any pump will not have adverse effects on the mains and other users.

2.3 Storage cisterns

Drinking water storage cisterns and lids, which include all cisterns used for domestic purposes, should not impart taste, colour, odour or toxicity to the water, nor promote microbial growth.

Cisterns for use with cold water should be listed in the *Water Fittings and Materials Directory* or conform to one of the following British Standards specifications:

- BS 417 Galvanized low carbon steel cisterns, cistern lids, tanks and cylinders. Part 2: Metric units.
- BS 1563 Cast iron sectional tanks (rectangular).
- BS 1564 Pressed steel sectional rectangular tanks.
- BS 1565 Specification for galvanized mild steel indirect cylinders, annular or saddle-back type.
- BS 4213 Specification for cold water storage and combined feed and expansion cisterns (polyolefin or olefin copolymer) up to 500 l capacity used for domestic purposes.

Cistern requirements

The following cistern requirements are noted (see figure 2.5):

- Materials should be suitable for maintaining potable water quality and must not deform unduly in use.
- All cisterns and pipes should be insulated against the effects of frost or heat.
- Cisterns should be situated away from heat.
- Access should be provided for inspection and maintenance (both internally and externally).
- All domestic cisterns and those supplying drinking water should be 'protected' to prevent contamination of water and should comply with the requirements of Water Regulations.
- Lids to be rigid, close fitting and securely fixed, and should fit closely around any vent pipe.

BS 6700 specifies that water cisterns for domestic purposes must be of the 'protected' type, on the grounds that water is likely to be drunk from all taps in dwellings. This is a departure from past practice which should greatly improve cisterns hygienically.

All cisterns should be supported on a firm level base capable of withstanding the weight of the cistern when filled with water to the rim. Where cisterns are located in the roof space, the load should be spread over as many joists as possible (see figure 2.6).

Occasionally large cisterns are buried or sunk in the ground. In these cases measures need to be taken to detect leakage and to protect the cistern from contamination (see figure 2.7).

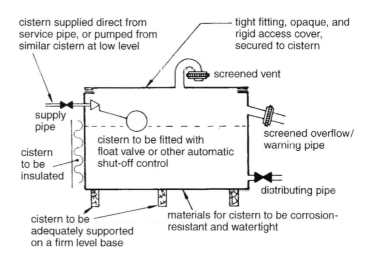

cistern supplied direct from service pipe, or pumped from similar cistern at low level

tight fitting, opaque, and rigid access cover, secured to cistern

screened vent

supply pipe

cistern to be fitted with float valve or other automatic shut-off control

screened overflow/ warning pipe

cistern to be insulated

distributing pipe

cistern to be adequately supported on a firm level base

materials for cistern to be corrosion-resistant and watertight

(a) General requirements for cisterns

Figure 2.5   Cistern requirements

continued

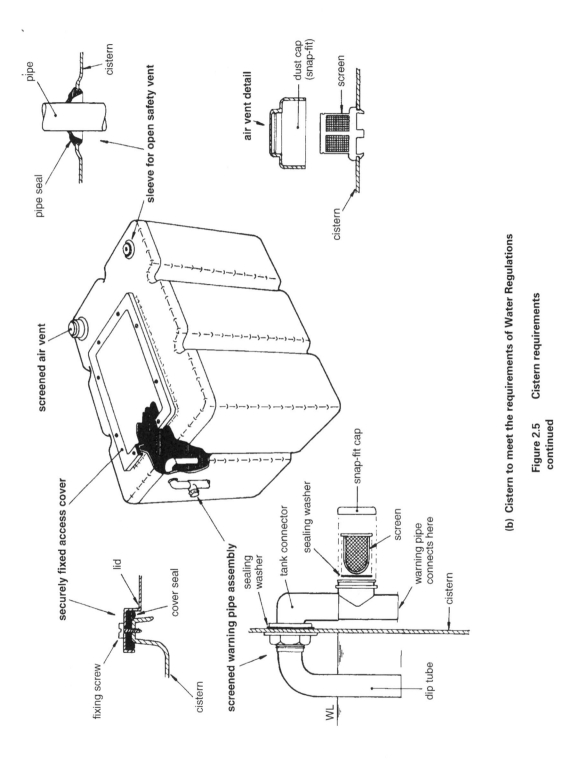

pipe

cistern

pipe seal

sleeve for open safety vent

air vent detail

dust cap (snap-fit)

screen

cistern

screened air vent

securely fixed access cover

screened warning pipe assembly

lid

cover seal

fixing screw

cistern

sealing washer

tank connector

sealing washer

snap-fit cap

screen

warning pipe connects here

cistern

dip tube

WL

(b) Cistern to meet the requirements of Water Regulations

Figure 2.5   **Cistern requirements**
continued

(a) Flexible cisterns

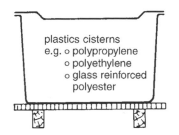

plastics cisterns
e.g. o polypropylene
     o polyethylene
     o glass reinforced
       polyester

Continuous support needed over whole base area.

No connections to be made to base of plastics cisterns.

(b) Rigid cisterns

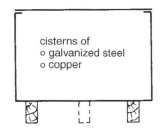

cisterns of
o galvanized steel
o copper

Two or more timber joists depending on size of cistern.

Continuous support not needed, and is undesirable for galvanized steel cisterns.

Figure 2.6   Support for cisterns

The installation should include:
- an electrical sump pump to automatically remove any water that might build up in the chamber due to the cistern overflowing, or through ingress of surface water;
- an audible or visual warning device to indicate when the cistern water reaches overflowing level;
- an audible or visual warning to show if cistern water or surface water are building up in the chamber.

Depending on ground conditions, cisterns may need to be anchored to prevent them lifting if the chamber becomes flooded and the cistern is empty or partially full.

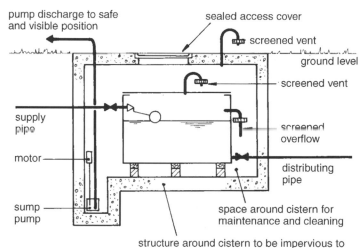

Figure 2.7   Sunken cistern

Cistern control valves

Every pipe supplying water to a cistern should be fitted with a float-operated valve or some other equally effective device to control the inflow of water and to maintain it at the required level.

Float valves should comply with BS 1212: Parts 1 to 4 (see figures 2.8, 2.9 and 2.11) and be used with a float complying with BS 1968 or BS 2456 of the correct size corresponding to the length of the lever arm and the water supply pressure.

Float valves should be clearly marked with the water pressure, temperature and other characteristics for which they are intended to be used.

Every float-operated valve should be securely fixed to the cistern and,

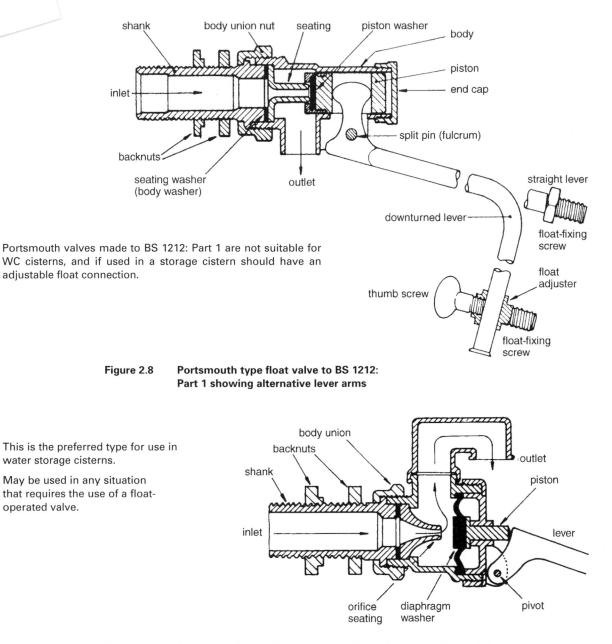

Portsmouth valves made to BS 1212: Part 1 are not suitable for WC cisterns, and if used in a storage cistern should have an adjustable float connection.

Figure 2.8   **Portsmouth type float valve to BS 1212: Part 1 showing alternative lever arms**

This is the preferred type for use in water storage cisterns.

May be used in any situation that requires the use of a float-operated valve.

Figure 2.9   **Diaphragm float valve to BS 1212: Part 2 (brass) and Part 3 (plastics)**

where necessary, braced to prevent any movement of the float or cistern wall which might affect the inflow to the cistern, or cause noise.

Where a non-British Standard float-operated valve or other level control device is used, it must meet the requirements of Water Regulations and be listed in the *Water Fittings and Materials Directory* produced by the Water Regulations Advisory Scheme (WRAS). Figures 2.10 and 2.12 illustrate two types of non-British standard valves.

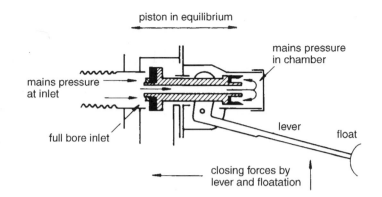

Figure 2.10   **Portsmouth equilibrium float valve**

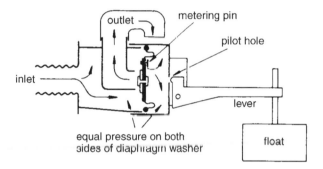

BS 1212: Part 4 covers this type of valve which is designed
primarily for use in WC flushing cisterns.

Figure 2.11   **Diaphragm equilibrium float valve**

Used to provide a full flow of water at all times, it completely eliminates 'dribble conditions' often associated with a conventional ball valve; this valve is particularly applicable to automatic pumping and booster systems and water treatment. 'Arclion' is a registered trade name of H. Warner & Son Ltd.

(a) Cistern emptying

As water is drawn from the storage cistern, the canister valve remains closed, thus preventing the main float-operated valve from opening.

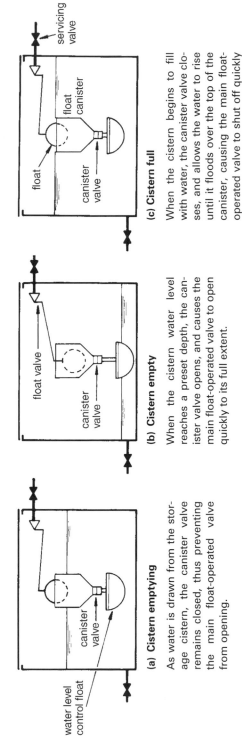

(b) Cistern empty

When the cistern water level reaches a preset depth, the canister valve opens, and causes the main float-operated valve to open quickly to its full extent.

(c) Cistern full

When the cistern begins to fill with water, the canister valve closes, and allows the water to rise until it floods over the top of the canister, causing the main float-operated valve to shut off quickly and cleanly.

Overflow/warning pipes not shown.

Figure 2.12   **Arclion® Delayed action float valve**

Connections to cisterns

Connections to cisterns will typically be similar to that shown in figures 2.13 and 2.14. It is advised that all connections be made at prescribed distances above the base of cisterns to permit debris particles to settle out rather than pass into the water system and possibly cause problems in tap seatings or other fittings with moving parts.

Warning/overflow pipe must be large enough to carry away leakage under worst conditions.

Dimensions are in millimetres.

Cistern must comply with Water Regulations if it is used for domestic purposes.

Float valve connection should be braced to prevent movement by water pressure thrust.

Dimension 'A'. Distributing pipe connection should be as low as is practically possible.

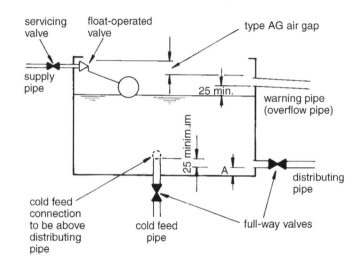

(a) Small cisterns (up to 1000 l capacity)

Washout connection flush with base of cistern at its lowest point.

Washout pipe to be plugged when not in use and must discharge to open air at least 150 mm above any drain.

Overflow pipe must be large enough to carry away leakage under worst conditions.

Cold feed and cold distributing pipes to be fitted with corrosion-resistant strainers.

Cold feed pipe to be above distributing pipe.

Dimensions are in millimetres.

Cistern must comply with Water Regulations if it is used for domestic purposes.

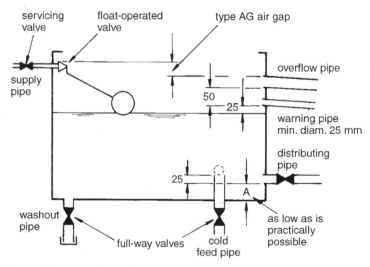

(b) Large cisterns (over 1000 l capacity)

Figure 2.13   **Connections to cisterns**

It is thought that to reduce the risk of contamination with *Legionella* and other similar water-borne diseases, outlet connections should be positioned as near to the bottom of the cistern as is reasonably practical to enable small particles to pass through the system rather than to cause unhealthy accumulations at the base of cisterns. This should not, however, encourage connections of pipes to the underside of any cistern, particularly those of flexible materials that need to be fully supported over the whole base area.

Outlet connections to hot water apparatus, for example cold feed pipes, should be positioned 25 mm above connections to cold water fittings. This is especially important where mixer fittings and showers are fitted and will, if the storage cistern should fail, enable the hot water to shut off before the cold and thus reduce the risk of scalding.

Linked cisterns

On occasion, to provide large quantities of water storage, or because of space restrictions, two or more cold water storage cisterns may need to be linked. Figure 2.14 shows two methods which permit cisterns to be cleaned, repaired or replaced without interruption of supplies to the building.

To avoid *Legionella* cisterns should:
- be small enough to ensure a rapid turnover and thus prevent stagnation;
- have float valves arraged to open and close together;
- have inlet and outlet connections at opposite ends;
- be regularly inspected and maintained in clean condition;
- conform to the requirements of Water Regulations.

The diagram shows separate overflow pipes. A common overflow pipe or warning pipe may be fitted provided cisterns are linked to form one storage unit.

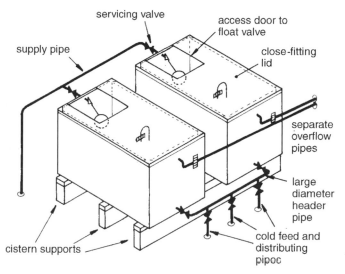

(a) Cisterns connected in parallel

This method may be preferred for prevention of *Legionella*.

To take cistern 1 out of commission for cleansing:
- fit temporary connection between link (b) to distributing pipe;
- removo link (a) and cap off close to cisterns.

To take cistern 2 out of commission for cleansing
- fit temporary connection to float valve in cistern 1 and disconnect branch pipe to float valve in cistern 2;
- remove links (a) and (b) and cap off near cisterns.

Sterilize any pipes and fittings used before they are fitted.

Sterilize cistern and pipes before putting back into service.

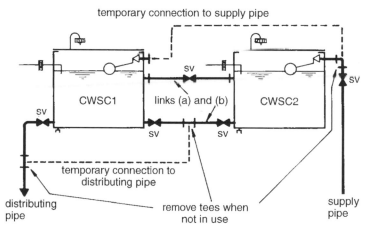

(b) Cisterns connected in series

Figure 2.14   Linked cisterns

Cisterns mounted outside buildings

Whether fixed to the building itself or supported on an independent structure, cisterns outside buildings, see figure 2.15, should be enclosed in a well-ventilated, yet draughtproof housing. This should be constructed to prevent the entry of birds, animals and insects, but provide access to the interior of the cistern for maintenance. Ventilation openings should be screened by a corrosion-resistant mesh with a maximum aperture size of 0.65 mm.

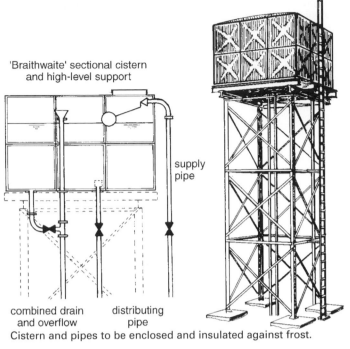

'Braithwaite' sectional cistern
and high-level support

supply
pipe

Washout fitted flush with base of
cistern, and base of cistern laid to
fall to asssist draining and cleanup.

combined drain   distributing
and overflow   pipe

Cistern and pipes to be enclosed and insulated against frost.

Enclosure to be ventilated, but draughtproof, and arranged to
prevent entry by birds, animals and insects.

Overflow/drain pipe must terminate in a conspicuous position
above ground level.

Figure 2.15   **Typical exterior storage cistern**

Large cisterns

Generally these provide over 5000 l storage and are often made up of
preformed panels or are constructed of concrete. They should preferably
be divided into two or more compartments to avoid interruption of the
water supply when carrying out repairs or maintenance to the cistern.

Cistern capacities

Commentary to clause 2.2.3.1.1 of BS 6700 makes the following
recommendations for houses:

smaller houses	cold water outlets only	100 l to 150 l	
	hot and cold outlets	200 l to 300 l	
larger houses	per bedroom	80 l	where adequate supply pressures are guaranteed
		130 l	where refilling only takes place at night

See section 5.3 for more detailed cistern sizing.

Pipework to and from cisterns

Supply and distributing pipes to and from cisterns should comply with the following recommendations.

(1)   All pipework should be insulated to reduce heat losses and gains, to minimize frost damage, and to prevent condensation.
(2)   Pipes should preferably be laid to a fall to reduce the risk of air locks and to facilitate filling and draining.
(3)   Pipes should be closely grouped for neatness, but not so close as to gain heat from one another.
(4)   Pipes should be securely fixed and adequately supported.
(5)   A cold feed pipe should not be used other than to supply the hot water apparatus for which it is intended.

Warning and overflow pipes

Warning and overflow pipes (see figure 2.16) serve two purposes:

(1)   to give warning that inlet valve to cistern has failed to close,
(2)   to remove safely from the buildings any water which leaks from the inflow pipe.

On cisterns of less than 1000 l one pipe only will serve both as an overflow pipe and a warning pipe, but on larger cisterns it may be necessary to have a separate pipe for each function (see figures 2.16–2.18).

Note   For Water Regulations purposes cistern capacities are measured to the level at which the water starts to flow into the overflow pipe.

The following recommendations for warning pipes and overflow pipes are noted (see figures 2.19–2.21).

(1)   Overflow and warning pipes should be made of rigid, corrosion-resistant material.
(2)   Overflow and warning pipes fitted to feed and expansion cisterns must be of metal, cross-linked polyethylene, polybutylene or chlorinated PVC, and be able to resist heat.
(3)   The overflow pipe or pipes should be able to carry away all water which is discharged into the cistern in the event of the inlet control becoming defective, without the water level reaching the spill-over level of the cistern, or submerging the discharge orifice of the inlet pipe or valve. Every overflow pipe shall discharge immediately the water in the cistern reaches the overflowing level (invert of overflow pipe).
(4)   Warning and overflow pipes must fall continuously to points of discharge.

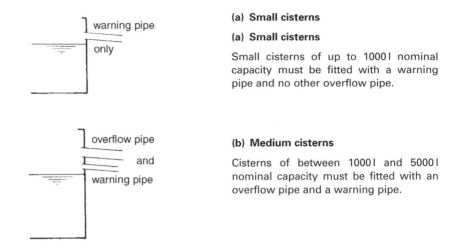

(a) Small cisterns

Small cisterns of up to 1000 l nominal capacity must be fitted with a warning pipe and no other overflow pipe.

(b) Medium cisterns

Cisterns of between 1000 l and 5000 l nominal capacity must be fitted with an overflow pipe and a warning pipe.

Cisterns of more than 5000 l nominal capacity may have other warning devices fitted in place of the warning pipe (see figures 2.17 and 2.18).

Figure 2.16   **Warning and overflow pipes for cisterns**

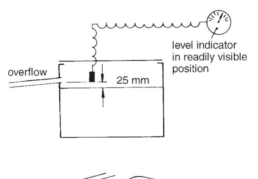

(a) Electrically operated warning device

Cisterns of between 5000 l and 10 000 l nominal capacity must be fitted with an overflow pipe and a warning pipe. Alternatively the warning pipe may be replaced by a level indicator that will clearly show when the water level is 25 mm below the overflowing level.

(b) Float-operated warning device

For use with cisterns of between 5000 l and 10 000 l nominal capacity.

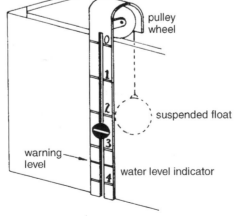
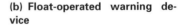

Figure 2.17   **Overflow and warning arrangements for large cisterns**

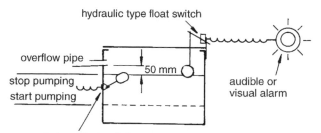

(a) Electrically operated alarm

Cisterns of more than 10 000 l nominal capacity must be fitted with an overflow pipe and a warning pipe. Alternatively the warning pipe may be replaced by an audible or visual alarm that will clearly show when the water level is 50 mm below the overflowing level.

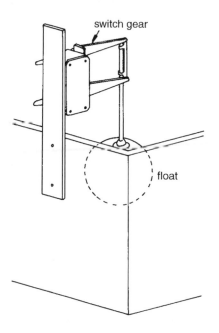

(b) Hydraulic type float switch

Float switch can be used to operate an audible or visual alarm. Also used for water level control.

Figure 2.18   Overflow and warning arrangements for very large cisterns

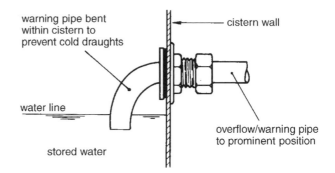

warning pipe bent
within cistern to
prevent cold draughts

cistern wall

water line

overflow/warning pipe
to prominent position

stored water

Figure 2.19   Overflow/warning pipe

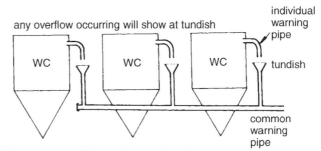

any overflow occurring will show at tundish

individual
warning
pipe

WC   WC   WC

tundish

common
warning
pipe

Cisterns must all be at the same level.

Individual warning pipes must terminate over tundish before connection to common pipe, so that any discharge is readily visible.

Common warning pipe must terminate in a conspicuous position.

Figure 2.20   Common warning pipes to WCs

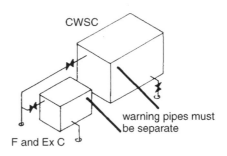

CWSC

warning pipes must
be separate

F and Ex C

Figure 2.21   Warning pipes from feed and expansion cisterns

(5)   Where overflow and warning pipes discharge through the external wall of a building they should be arranged to prevent cold draughts by turning down the warning pipe into the cistern and below the water line (see figure 2.19).

(6)   Every warning pipe should discharge in a conspicuous position, preferably outside the building where this is appropriate. In some circumstances warning pipes may be permitted to terminate into a special bath overflow fitting.

(7)   Warning pipes from more than one similar cistern may be linked together to one common outlet providing they cannot discharge one into the other, and as long as the source of any overflow can be readily identified (see figure 2.20).

(8)   Warning pipes from feed and expansion cisterns must be separate from those from storage cisterns.

(9)   When a single overflow pipe or warning pipe is used, it should have a minimum diameter of 19 mm, or be at least one size larger than the cistern inlet pipe, whichever is the greater.

2.4   Valves and controls

Stopvalves, servicing valves and isolating valves

Water Regulations require the use of three types of control valve on water installations: stopvalves, servicing valves and isolating valves.

Stopvalves

A supply stopvalve must be provided on the supply pipe, inside the building, to control the whole of the water supply to the building. It should be in an accessible position above floor level, and near the point of entry of the pipe. Suitable types of stop valve are illustrated in figure 2.22.

Stopvalves to BS 1010 are for above ground use only.

Stopvalves to BS 5433 for underground use are of heavier quality and made of corrosion-resistant material such as gunmetal or DZR brass and have an enclosed (shrouded) washer.

Washer plates should be 'fixed' so as to open mechanically when valve spindle is screwed to open position.

Illustration shows stopvalve to BS 1010 with type A compression joint for copper.

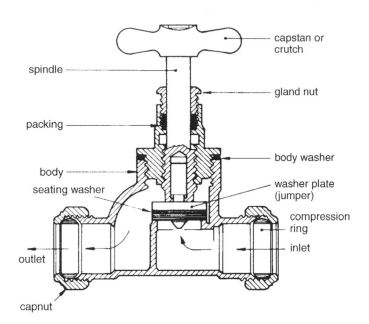

(a) Screwdown stopvalve to BS 1010 and BS 5433

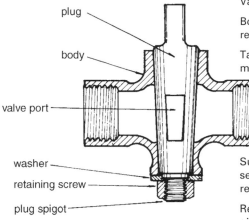

Valve shown in closed position.

Body of cast gunmetal or other corrosion-resisting material such as DZR brass.

Tapered plug and valve body are machined to make a watertight joint.

Suitable non-toxic lubrication applied to seating surfaces of plug to ensure smooth revolving action.

Retaining screw used to adjust plug to give watertight seal but allow easy shut-off movement.

Open area of valve port to coincide with waterway in valve body to give full flow with minimum resistance.

Valve connections shown are for threated joint to BS 21.

Valves are available with ends for other materials, i.e. copper, polyethylene.

(b) Underground plug cock to BS 2580

Figure 2.22   Stopvalves

continued

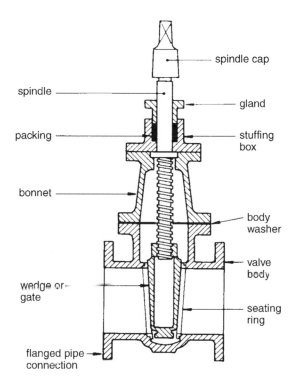

spindle cap

spindle

gland

packing

stuffing box

bonnet

body washer

valve body

wedge or gate

seating ring

flanged pipe connection

(c) Flanged gate valve to BS 5163

Figure 2.22 **Stopvalves**
continued

Stopvalve positions and supply pipe arrangements

It is advisable that every separate premises or building that is supplied with water, be fitted with a separate isolating stopvalve close to its branch connection so that the whole of each branch pipe is fully controlled.

It should be possible to turn off the supply to premises in an emergency or for repair without affecting the supply to any other premises.

There are various methods of supply to premises that allow proper control of supply pipes and these are shown in figures 2.23–2.25.

It should be borne in mind that the water undertaker may not permit the use of joint supply pipes. If these are planned the water undertaker's advice should be sought.

When a stopvalve is installed on an underground pipe it should be enclosed in a pipe guard under a surface box (see figures 10.5 and 11.57–11.60).

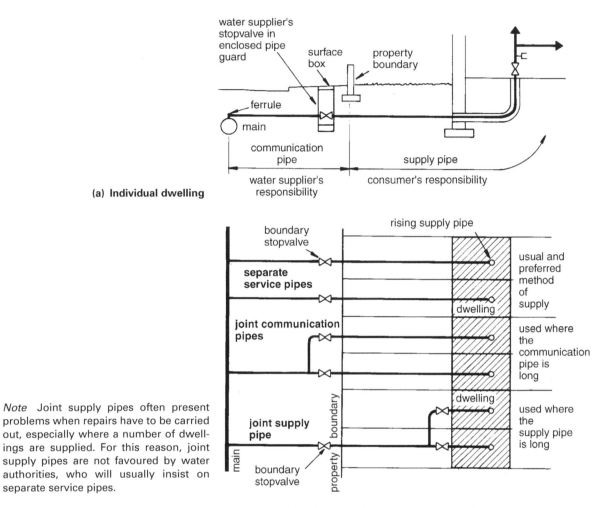

(a) Individual dwelling

Note Joint supply pipes often present problems when repairs have to be carried out, especially where a number of dwellings are supplied. For this reason, joint supply pipes are not favoured by water authorities, who will usually insist on separate service pipes.

(b) Methods of supplying more than one dwelling

Figure 2.23   Stopvalves to dwellings

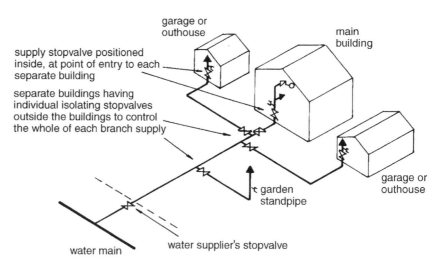

supply stopvalve positioned
inside, at point of entry to each
separate building

separate buildings having
individual isolating stopvalves
outside the buildings to control
the whole of each branch supply

garage or
outhouse

main
building

garage or
outhouse

garden
standpipe

water main

water supplier's stopvalve

Figure 2.24   **Stopvalves for premises having separate buildings**

A similar arrangement may be necessary
where separate premises are supplied
from a common distributing pipe from a
common cold water storage cistern at
high level (see figure 6.23b).

Backflow protection may be required at
each floor level.

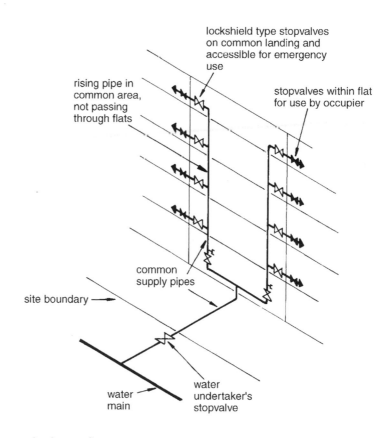

lockshield type stopvalves
on common landing and
accessible for emergency
use

rising pipe in
common area,
not passing
through flats

stopvalves within flat
for use by occupier

common
supply pipes

site boundary

water
main

water
undertaker's
stopvalve

Figure 2.25   **Common supply pipes to flats**

continued

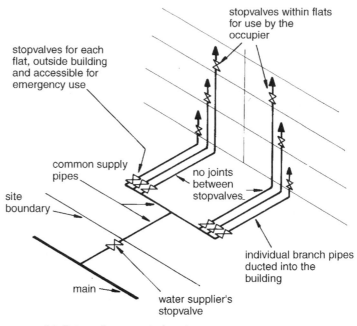

stopvalves within flats
for use by the
occupier

stopvalves for each
flat, outside building
and accessible for
emergency use

common supply
pipes

no joints
between
stopvalves

site
boundary

individual branch pipes
ducted into the
building

main

water supplier's
stopvalve

(b) Externally separated system

The externally separated system is the method preferred by
water authorities who may also insist on a separate mains connection for each flat

Figure 2.25   **Common supply pipes to flats**
continued

Servicing valves

Servicing valves must be provided in accessible positions to enable the flow
of water to individual or groups of appliances to be controlled and to limit
inconvenience caused during maintenance and repair (see figures 2.26 and
2.27).

Traditionally gate valves have been used for this purpose rather than
screwdown valves because of their lower hydraulic resistance. In recent
years spherical plug valves have been introduced and are particularly suited
to control individual fittings or appliances.

Servicing valves must be fitted:

- immediately before every float valve connected to a service pipe or
  distributing pipe;
- to every pipe carrying water from a hot water storage cylinder, tank or
  cistern;
- on every cold feed and distributing pipe from any feed cistern or storage
  cistern of more than 18 l capacity.

However, cold feed to primary hot water circuits should not be fitted with a
servicing valve.

Stopvalves

Stopvalves that meet the requirements of Water Regulations.

Up to 50 mm diameter:
- ○ screwdown stopvalves to BS 5433
- ○ plug cock to BS 2580
- ○ screwdown stopvalves to BS 1010 (above ground use only)

Above 50 mm diameter:
- ○ flanged gate valve to BS 5163

Servicing valves and isolating valves

- ○ any of the stopvalves listed above, or
- ○ wheel operated (gate) valve to BS 5154
- ○ lever-operated, spherical plug valve to BS 6675
- ○ Screwdriver-operated (slot type), spherical plug valve to BS 6675

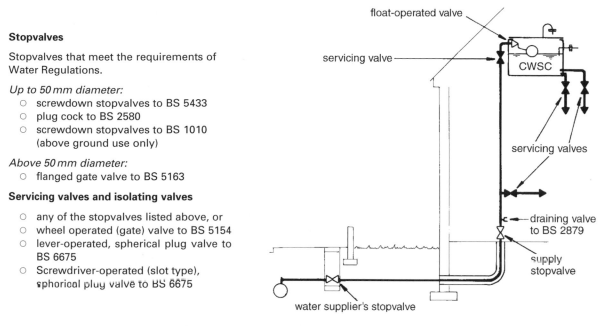

Figure 2.26   Control valves within dwellings

Isolating Valves

Isolating valves are required by Water Regulations 'to be installed to isolate parts of the system'. These will be useful for maintenance and for closing down parts of the supply when leakages are taking place.

On larger systems control valves should be fitted:

- • to isolate pipework on different floors;
- • to isolate various parts of an installation at the same level;
- • to isolate branch pipes that lead to a range of appliances.

(a) Gate valve to BS 5154

Can be used on distributing pipe outlets from cisterns.

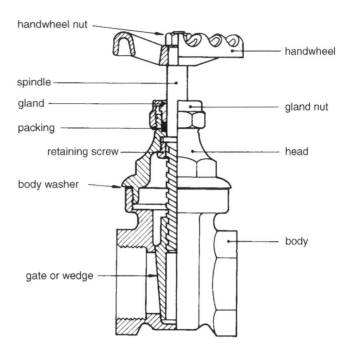

(b) Spherical plug valves to BS 6675

Suitable for use near to single outlet fittings and appliances.

Available in sizes up to 25 mm diameter.

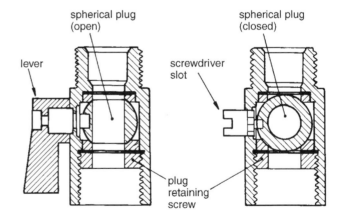

Figure 2.27   Types of servicing valve

Stop valves and draining valves

Water Regulations require that 'sufficient number of stopvalves and draining valves be fitted to minimize the discharge of water when water fittings are maintained or replaced'.

Draining valves should be fitted at all low points on supply pipes and distributing pipes so that systems can be drained for frost protection and for convenience when carrying out repairs.

2.5   Water revenue meters

This section should be regarded as for information only.

Meters may be fitted by the consumer's contractor or by the water undertaker. Where fitting is done by a private contractor the water supplier will need to be consulted and agreement reached on siting and installation details.

The revenue meter will be supplied by, and will remain the property of, the water supplier.

The preferred position for the meter is at the boundary of the premises at the end of the communication pipe so that it will register the whole supply (see figure 2.28).

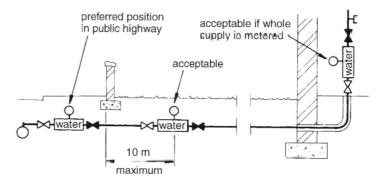

Figure 2.20   Meter positions

In the case of premises with multiple occupations, e.g. flats, and where underground installations are not practicable, an internal installation may be acceptable provided the whole supply is registered.

Notes on fitting

The following points on fitting should be noted:

- The meter installation should comply with BS 5728.
- The meter must be protected from risk of damage by shock or vibration.
- The method of connection should permit meter changes without the use of heat or major disturbance of pipework.
- Connector unions should have flat seats and be complete with 1.5 mm thick washers to allow for tolerance on meter lengths.
- Meter lengths
  - G3/4B meters, 134 mm;
  - G1B meters, see BS 5728.

Flow rates

Nominal flow rate for the G3/4B meter is $1.5\,m^3/h$, and for the G1B meter is $2.5\,m^3/h$.

Continuity bonding

BS 7671 (Requirements for Electrical Installations – IEE Wiring Regulations) does not allow the use of water pipes as an electrode for earthing purposes. However, any metal water supply pipe must be bonded to the electrical installation main earth terminal as near as possible to its point of entry into the building.

A suitable conductor should be installed between pipes on either side of the meter and stopvalves, to protect the installer against electrical fault, and for the maintenance of the earth connection during use, and particularly when the meter is being replaced.

For dwellings the bond should have a cross-sectional area of at least $6\,mm^2$.

External meter installations

- Meters should be positioned below the ground, in a suitable chamber that will permit ample space for meter removal or replacement.
- For smaller installations of up to 3.5 l/h the chamber may be constructed of glass reinforced plastics or PVC; larger installations should have chambers of brick or concrete.
- The chamber should be well constructed, with a cover marked 'water meter' and fitted with slots or lifting eyes.
- The clear opening of the surface box (cover) should be equal to the internal dimensions of the chamber. Covers should not be of concrete.
- A meter below ground should be installed in a horizontal position.
- The meter should be fitted with isolating valves on both inlet and outlet to facilitate meter changing.
- The meter should be supported so as not to cause any differential load on its connecting pipework.
- Pipes should be flexible enough, with space around, to permit meter exchanges, but at the same time secured against movement of any mechanical joint.
- Pipes, cables and drains other than meter pipework must not pass through the meter chamber.

See table 2.1 and figure 2.29.

Note   Whilst BS 6700 and many water authorities favour raising the meter as shown in figure 2.29 to make meter reading and changing easier, the author would prefer the meter to remain at the same level as the service pipe, as in figure 2.30, in order to comply with the Water Regulations requirement of 750 mm depth to avoid frost damage.

Table 2.1 **Recommended internal dimensions for meter chambers**

Size of meter	Size of chamber			Remarks
	Length	Breadth	Depth	
15 to 20	430	280	To suit	Or 380 circular
25	600	600	pattern	
40 to 50	900	600	of meter	
80*	1900	750	but not	
100*	2000	750	less than	
150*	2150	750	750	

* Dimensions for these sizes take account of compound assemblies but do not provide for isolating valves

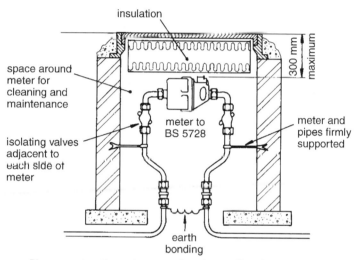

Pipes passing through meter chamber wall or floor should have clearance space to permit changing of meter and to prevent damage through ground movement.

Stopvalves shown are not in the most accessible positions.

Figure 2.29 **Below ground meter installation**

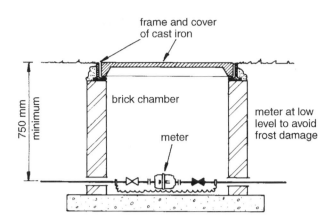

Figure 2.30 **Meter installation to avoid frost damage**

Internal meters

Internal meters may be fixed horizontally or vertically provided the dial is not more than 1.5 m above floor level and readily visible for reading (see figure 2.31).

Pipework to be adequately supported

Meter to be protected from frost, especially in exposed positions, e.g. garage.

Meter in cupboard may be brought forward to within 300 mm of the front of cupboard for ease of reading, providing it is properly supported.

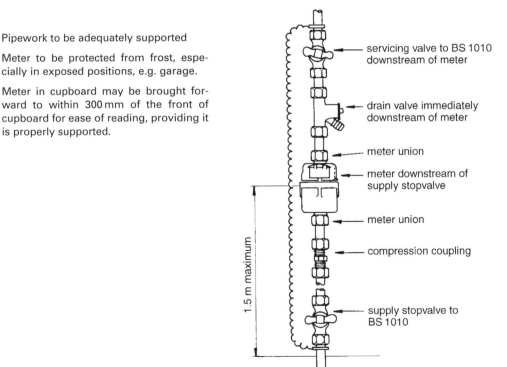

Figure 2.31 **Above ground meter installation**

Non-revenue meters

Installation should be as described for revenue meters but the water undertaker will not need to be consulted about the meter's position.

Patent meter connections

These save space and installation costs (see figure 2.32).

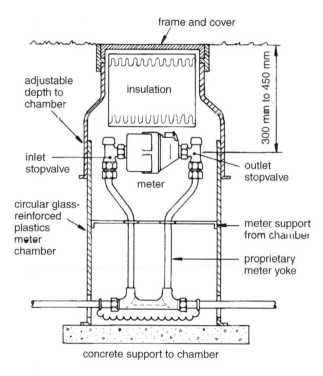

(a) Below ground, using meter yoke

Figure 2.32   **Patent meter connections**

continued

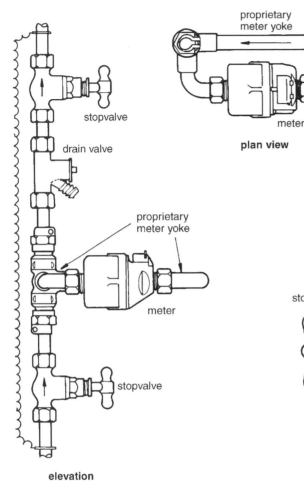

plan view

elevation

(b) Above ground, using meter yoke

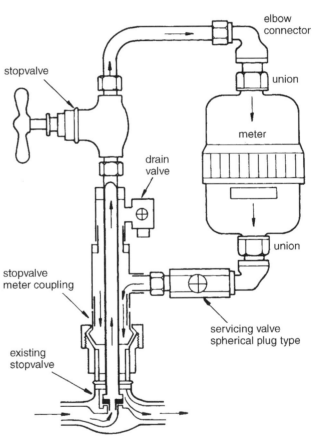

(c) Above ground, using stopvalve body for connection

Figure 2.32 **Patent meter connections**
continued

Meter readings

A typical domestic meter reading is shown in figure 2.33.

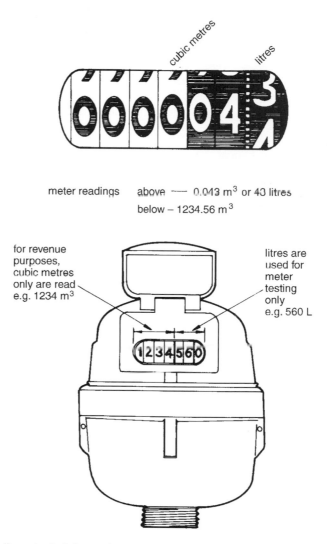

meter readings   above —— 0.043 m³ or 40 litres

below – 1234.56 m³

for revenue
purposes,
cubic metres
only are read
e.g. 1234 m³

litres are
used for
meter
testing
only
e.g. 560 L

Figure 2.33   Meter reading – typical domestic meter

2.6   Boosted systems

Water supplies to buildings vary greatly in pressure and quantity available. In some cases this may give rise to intermittent supplies and in others, especially high-rise developments, parts of the building may be above the pressure limit of the mains supply. In these situations there is a case for the use of pumps.

There are two ways that pumps can be used to deal with the above problems. These are by direct boosting and indirect boosting. Indirect systems are more common than direct systems; the latter are rarely permitted by water suppliers because they reduce the mains pressure available to other consumers and can increase the risks of backflow. Under no circumstances should any pump be connected directly to a supply pipe without first obtaining the written consent of the water undertaker.

Booster pumps can cause excessive aeration. Although this does not cause deterioration of water quality, the 'milky' appearance can cause concern amongst consumers.

It is desirable that sample taps are fitted at the outlet of pumps to provide for periodic sampling to ensure the continued quality of the pumped water.

Basic systems

The basic systems are as follows:

- simple direct boosting, see figure 2.34;
- direct boosting to header and duplicate storage cisterns, see figure 2.35;
- indirect boosting to storage cistern, see figure 2.36;
- indirect boosting with pressure vessel, see figure 2.37.

The *pneumatic pressure vessel*, as seen in figure 2.38, contains both compressed air and water. As the water is drawn off, the water level drops, the air expands with a resulting loss of water, and a float switch starts the pumps. The water level rises to a predetermined level at which the float switch stops the pumps. The sequence of events will continue to supply the building until the water level falls and the cycle begins again. As and when air is lost from the pressure vessel by absorption into the water, pressure will fall, and the air is replaced via the air compressor.

Some systems incorporate a sealed pressure vessel containing a flexible membrane to separate the air from the water. This will reduce air loss and eliminate the need for a permanent compressor. However, air pressures should be checked at intervals and replenished if necessary. Some units contain nitrogen instead of air, and it is important that these are topped up with nitrogen, not air or some other gas (see figure 2.39).

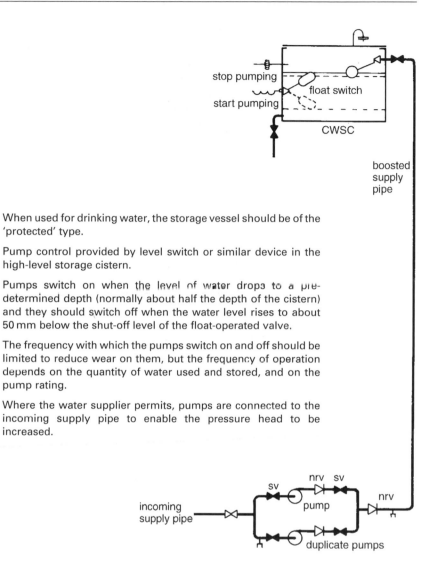

When used for drinking water, the storage vessel should be of the 'protected' type.

Pump control provided by level switch or similar device in the high-level storage cistern.

Pumps switch on when the level of water drops to a pre-determined depth (normally about half the depth of the cistern) and they should switch off when the water level rises to about 50 mm below the shut-off level of the float-operated valve.

The frequency with which the pumps switch on and off should be limited to reduce wear on them, but the frequency of operation depends on the quantity of water used and stored, and on the pump rating.

Where the water supplier permits, pumps are connected to the incoming supply pipe to enable the pressure head to be increased.

Figure 2.34   Simple direct boosting

System used for large and high-rise buildings.

Cisterns at high-level supply non-drinking water.

Float switches operate pump ON/OFF to control water level in cisterns.

Drinking water header provides limited storage for drinking water whilst pump is not running.

Drinking water header sized to provide 5 l to 7 l per dwelling per day.

Pipeline level switch on header bypass to start pumps when water level drops. Pumps can then be time controlled or arranged to shut off by pressure switch.

Drinking water supplies to sinks in flats taken from boosted supply pipe.

Pumps should be arranged to cut out approximately 50 mm below float valve shut-off level.

Secondary backflow devices may be required at each floor level.

Excessive pressure should not be allowed to build up or splashing at taps and increased waste of water may occur.

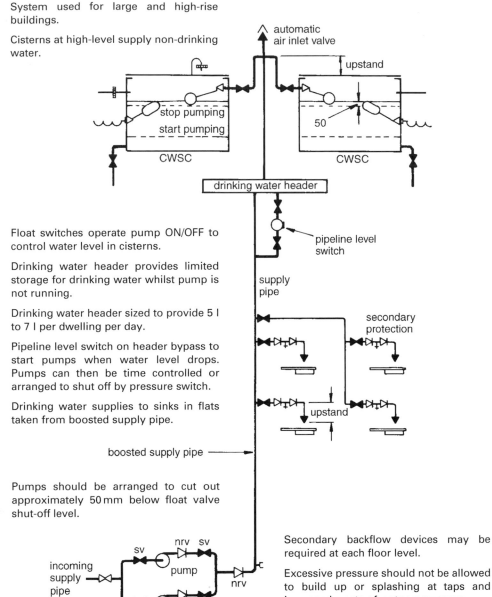

Figure 2.35   Direct boosting to header and duplicate storage cisterns

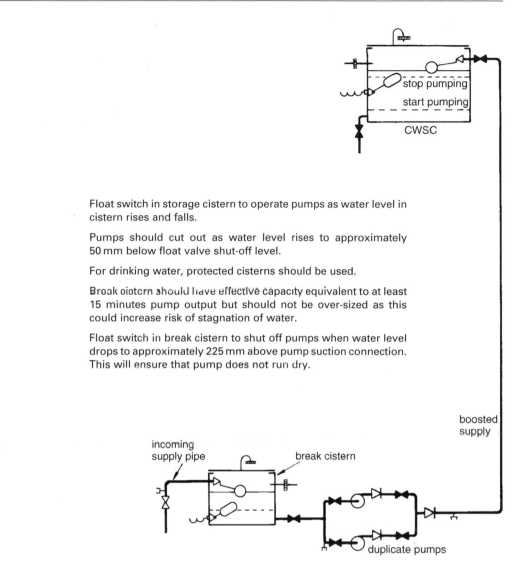

Float switch in storage cistern to operate pumps as water level in cistern rises and falls.

Pumps should cut out as water level rises to approximately 50 mm below float valve shut-off level.

For drinking water, protected cisterns should be used.

Break cistern should have effective capacity equivalent to at least 15 minutes pump output but should not be over-sized as this could increase risk of stagnation of water.

Float switch in break cistern to shut off pumps when water level drops to approximately 225 mm above pump suction connection. This will ensure that pump does not run dry.

Figure 2.36   **Indirect boosting to storage cistern**

For use in buildings where a number of storage cisterns are supplied at various levels and where it is not practicable to control pumps by level switches.

Normally pressure vessel, pumps, air compressor and control equipment are purchased as a packaged pressure set.

Floors above limit of mains pressure supplied via break cistern and pumps.

Unboosted supply to floors within limit of mains pressure.

Drinking water cisterns in dwellings must be protected from contamination.

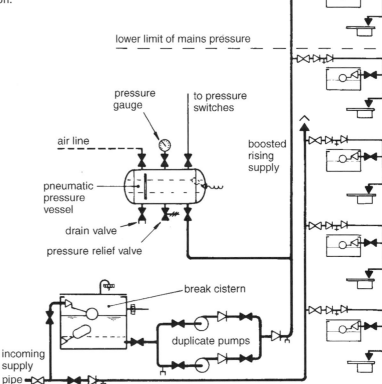

Secondary backflow protection is required at each floor level where separately chargeable premises are supplied.

Figure 2.37   Indirect boosting with pressure vessel

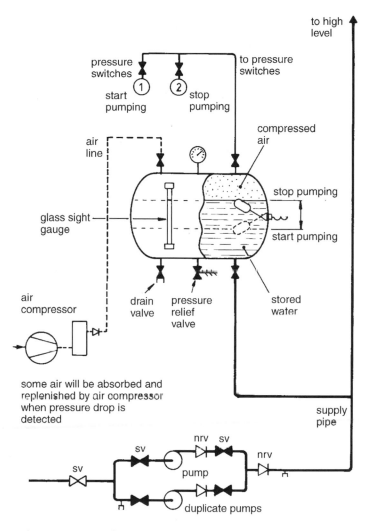

Figure 2.38   **Pneumatic pressure vessel**

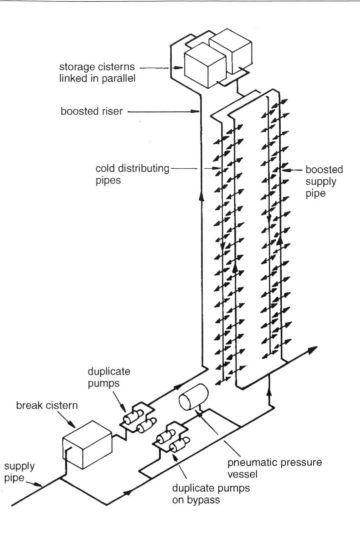

storage cisterns
linked in parallel

boosted riser

cold distributing
pipes

boosted
supply
pipe

duplicate
pumps

break cistern

supply
pipe

pneumatic pressure
vessel

duplicate pumps
on bypass

Drinking water supply direct from mains using pneumatic pressure vessel.

Indirect boosting for storage cisterns.

Control valves and backflow prevention devices omitted for clarity.

Figure 2.39   Typical installation to multi-storey building

Pumps and equipment

Pumps and other associated equipment are usually located within the building being served, preferably as near as possible to the point of entry of the incoming pipe.

Electrically driven centrifugal pumping plant is normally used; pumps and other equipment should be duplicated.

Pumps may be either horizontal or vertical types, directly coupled to their electric motors. A solid foundation is essential for all motors and pumps and anti-vibration mountings should normally be specified.

Automatic control of pumping plant is essential and pressure switches, level switches, or high-level and low-level electrodes should give reliable control. Other methods of control, both mechanical and electrical, could be considered. Pumping should be controlled using a pump selector switch and an ON/OFF/AUTO control. Motor starters should incorporate overload protection.

Pumps should be installed in duplicate and sized so that each pump is capable of overcoming the static lift plus the friction losses in both pipework and valves. Where pumps are connected directly to the service pipe, allowance should be made for the minimum operating pressure in the service pipe, since the pump head is added to this and does not cancel out any existing pressure.

Care should be taken in pump and pipe sizing to minimize the risk of water hammer due to surge when pumps are started and stopped.

Transmission of pump and motor noise via pipework can be reduced by the use of flexible connections. Small-power motors of the squirrel cage induction type are suitable for most installations. Low-speed pumps are preferable to promote a long efficient life and reasonably quiet operation. The fitting of motors with sleeve type super-silent bearings should be considered for quiet running.

All pipework connections to and from pumps should be adequately supported and anchored against thrust to avoid stress on pump casings and to ensure proper alignment.

Most small air compressors used for charging pneumatic pressure vessels are of the reciprocating type, either air-cooled or water-cooled. A water-cooled after-cooler for the condensation and extraction of oil and moisture from the compressed air should be installed. The air to be compressed should be drawn from a clean cool source and should be protected from contamination. Check or non-return valves should offer a minimum of frictional loss and should be non-concussive.

The pump room (see figure 2.40) should be of adequate size to accommodate all the plant and also provide adequate space for maintenance and replacement of parts; it should be dry, ventilated and protected from frost and flooding. Entry of birds and small animals must be prevented. Access should be restricted to authorized persons.

Provision should be made for the pumps to be supplied by an alternative electricity supply in the event of mains failure.

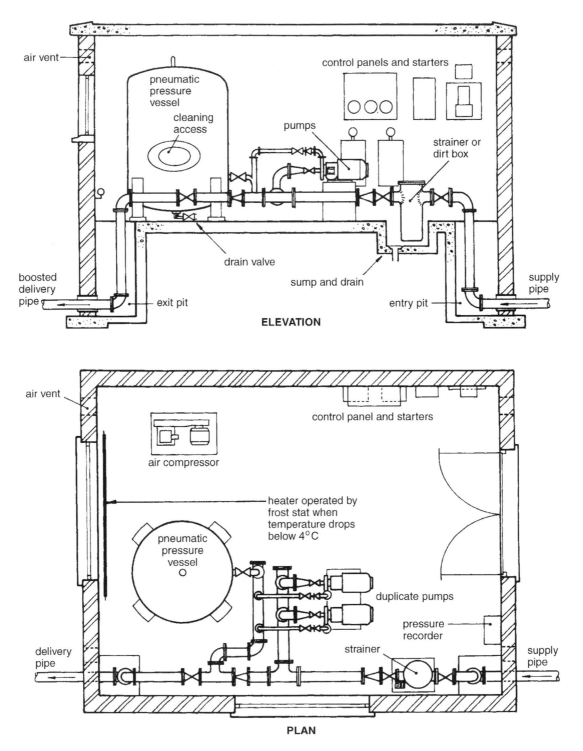

Figure 2.40   Typical pump room

Maintenance and inspection

A responsible person should be appointed to oversee the proper install-ation of any pumped water scheme, and the user should arrange for regular maintenance and inspection of the pumps and plant.

All work carried out and inspections made should be recorded in a suitable log book which should be kept in the plant room.

2.7 Water treatment

Water softeners

The main purpose of a water softener is to reduce scale formation in hot water systems and components. However, over-softening can increase the ability of the water to dissolve metals. This can be of particular concern where the water system contains lead pipes.

Water softeners are considered only briefly in BS 6700, its main con-cern being the risk to health should backflow occur and the protection that should be provided to prevent backflow from taking place.

Base exchange (ion exchange) water softeners work by passing the hard water through an enclosed tank containing resin particles (beads). The resin attracts and absorbs the hardness salts, mainly calcium and mag-nesium, from the water, and at the same time replaces them with sodium from the resin. After a while the resin becomes saturated with hardness salts and needs to be regenerated using a brine (salt) solution to put sodium back into the resin. The hardness salts are given up from the resin and washed to waste down the drain.

Modern water softeners are electronically controlled to recharge the resin, either at timed intervals or as the resin becomes saturated, and to automatically flush the residual hardness salt to waste. The only main-tenance requirement is the occasional replacement of salt in the brine tank (see figure 2.41).

Advantages

Some of the advantages of water softeners are as follows:

(1) Savings in soap and reduction of scum, resulting in:
   - reduced expenditure on soap purchase (small savings);
   - easier cleaning of appliances;
   - cleaner crockery from dishwashers.
(2) Smooth, gentle feel of bath and shower water.
(3) Scale reduction in appliances and components, resulting in:
   - longer life for cylinders, immersion heaters and other com-ponents;
   - less maintenance, e.g. shower outlets less likely to become clog-ged with fur.

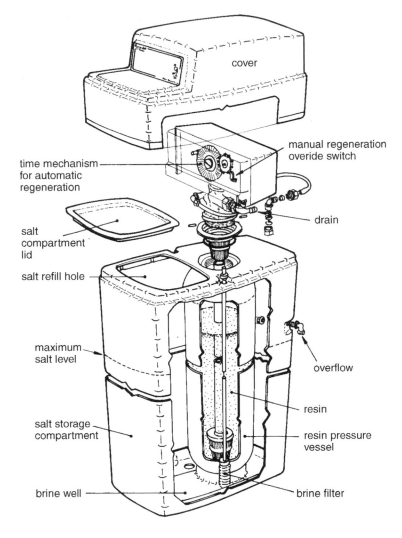

Figure 2.41   **Water softener, base exchange type**

Disadvantages

(1)   Additional installation costs (manufacturers claim a six-year payback).
(2)   Cost of running the unit, e.g. electricity supply and salt.
(3)   Drain needed for brine rinse.
(4)   User must add salt periodically.

Installation

Water softeners should be sited near the incoming supply pipe and where drain access is available (see figure 2.42).

Electricity supply is required for automatic control and operation.

Water softeners of the salt regenerated type installed in dwellings are

Backflow protection shown for domestic softener is a single check valve.

Non-domestic softeners require a double check valve assembly.

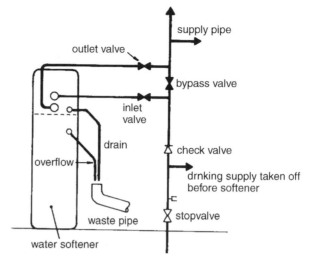

Figure 2.42   Installation of base exchange water softener for domestic use

considered to be a fluid category 3 backflow risk and as such a check valve should be installed on the supply pipe before the softener connection.

Drinking water supply should be taken off before the softener, and upstream of any check valves fitted.

In dwellings a single check valve is required under Water Regulations to protect the water supply from backflow.

A softener located other than in a dwelling requires the use of a double check valve arrangement or a combined check and anti-vacuum valve.

Other forms of treatment

As previously mentioned, BS 6700 deals briefly with water softeners, and does not refer to other forms of treatment which are shown in figures 2.43 and 2.44. It is not the intention here to recommend their use, rather to point out that there are such treatments readily available, and that they can in certain circumstances have a limited application.

Before any of the chemical devices are used, advice should be sought from local water undertakers, particularly regarding toxicity.

Polyphosphate dosing

This form of treatment can be used to reduce scale deposits, and in certain conditions can help protect pipes from corrosion. It usually consists of a small dosing chamber containing polyphosphate beads or crystals that dissolve as the water passes through (see figure 2.43).

In stagnant systems excessive concentrations of phosphates can build up to make the water less palatable, and could cause blockages due to the crystals reforming.

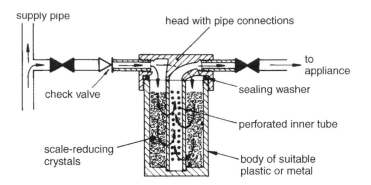

Useful for supplies to instantaneous water heaters.

Figure 2.43   **Pipeline dispenser**

Physical conditioners

Physical conditioners may be of a number of types, using a variety of methods to reduce the effects of scaling. These may be electronic, electrolytic, magnetic (figure 2.44), electromagnetic, or ionic devices that either have a direct water connection or are strapped in some way to the water pipe. The advantage of these is that they may give the effect of softened water without the addition or loss of mineral substances to the water. It should be said, however, that these devices do not soften the water, rather they inhibit the hardness particles so that they do not readily form scale deposits.

Physical conditioners are particularly suited to instantaneous type water heaters, but their effects may be lost in stored hot water.

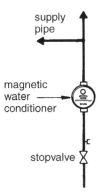

Precipitates the hardness salts into microscopic crystals by passing the water through a magnetic field.

Prevents scale forming as long as water is moving, but effects are said to lessen in stored water.

No backflow protection needed.

Figure 2.44   **Magnetic water conditioner**

Chapter 3
Hot water supply

Hot water is an essential requirement for all dwellings and most working environments. The following factors should be considered in the selection and the design of hot water supply systems:

(1) quantity of hot water required;
(2) temperature in storage and at outlets;
(3) cost of installation and maintenance;
(4) fuel energy requirements and running costs;
(5) waste of water and energy;
(6) safety of the user.

Hot water supply cannot be considered in isolation from central heating because systems commonly combine both functions. This book, like BS 6700, is primarily concerned with hot water supply and refers only to central heating in combined hot water and heating systems up to 44 kW output.

3.1 System choice

There are many methods by which hot water can be supplied, ranging from a simple gas or electric single point arrangement for one outlet, to the more complex centralized boiler systems supplying hot water to a number of outlets.

Figure 3.1, adapted from BS 6700, sets out a number of ways of supplying hot water, many of which are described in the following pages.

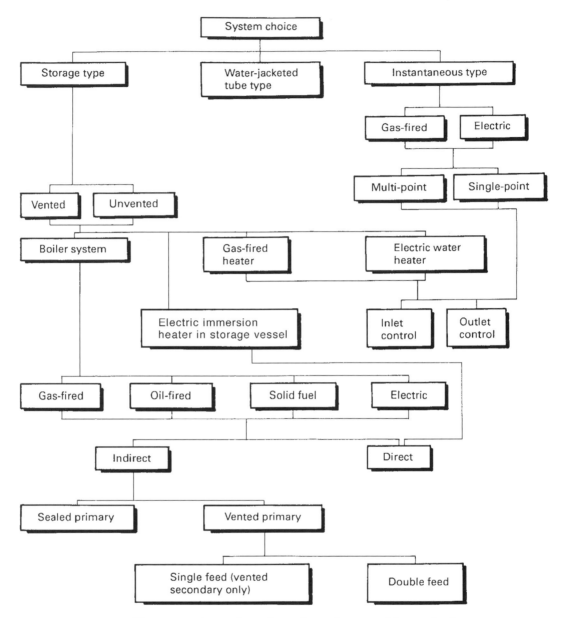

The system chosen depends on circumstances and the needs of the user, and may require the use of one method or a combination of two or more.

Figure 3.1   Alternatives for hot water supply

3.2   Instantaneous water heaters

Instantaneous water heaters (see figures 3.2 – 3.6) should be chosen with the following considerations in mind.

(1)   Some of these heaters have relatively high power ratings (up to 28 kW if gas fired) so it is important that adequate gas or electricity supplies are available.

(2)   The water in instantaneous water heaters is usually heated by about 55°C at its lowest flow rate, and its temperature will rise and fall inversely to its flow rate.

(3)   Where constant flow temperature is important, the heater should be fitted with a water governor at its inflow. Close control of temperature is of particular importance for showers.

(4)   To attain constant temperatures on delivery, water flow and pressure must also be constant. Variations in pressure can cause flow and temperature problems when the heater is in use, and when setting up or adjusting flow controls.

(5)   The use of multi-point heaters for showers should be avoided, except where the heater only feeds a bath with a shower over it.

(6)   Gas-fired heaters fitted in bathrooms must be of the room-sealed type. Room-sealed types are preferred in other locations.

(7)   Electrically powered heaters in bathrooms must be protected against the effects of steam and comply with BS 7671 (Requirements for Electrical Installations – IEE. Wiring Regulations).

Position heater near most-used appliance, usually the kitchen sink.

Where pipe runs from a multi-point heater are likely to be long consider using a number of single point heaters.

Multi-point heaters operate most satisfactorily when only one outlet is used at any one time.

Flow rate is variable and heater is generally not suited for use with a shower.

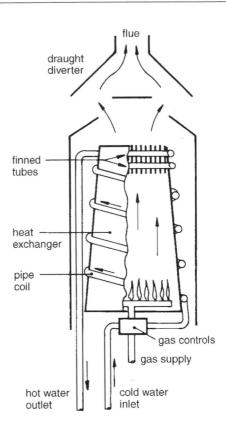

Figure 3.2   **Typical instantaneous water heater, gas-fired with conventional flue**

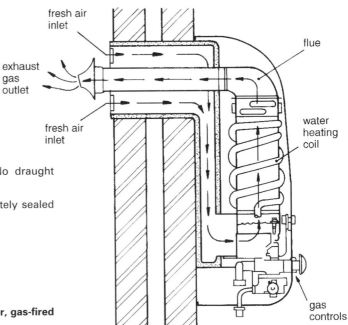

Flue inlet and outlet at equal air pressure. No draught problems to affect flame.

Combustion chamber, air inlet and flue completely sealed from room.

Figure 3.3   **Typical instantaeous water heater, gas-fired with balanced flue**

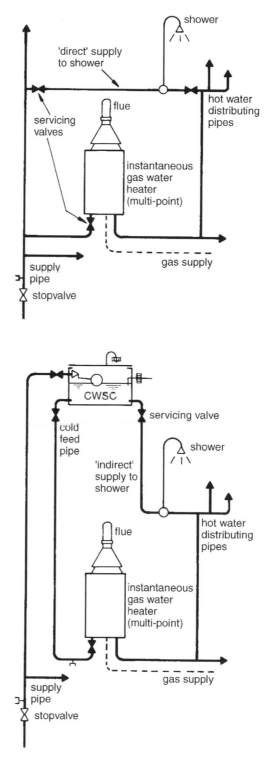

(a) Directly supplied heater

Constant flow rate needed to maintain 55°C temperature difference between feed water and heated water.

Pressure and flow variations will affect temperatures at outlets.

Showers are not recommended because of possible loss of constant temperature control and pressure.

Use only thermostatically controlled shower mixer.

The usual arrangement is direct from the supply pipe as shown here because installation cost is lower. However, supply from storage will give constant flow.

(b) Indirectly supplied heater

High installation cost compared with mains-fed system.

Constant pressure from storage for shower and other fittings gives more stable temperature control.

Figure 3.4   **Centralized gas-fired instantaneous water heater installations**

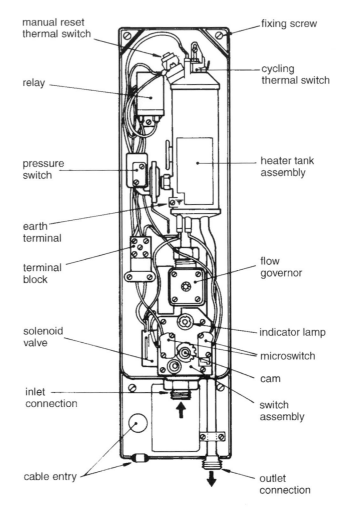

manual reset thermal switch

fixing screw

relay

cycling thermal switch

pressure switch

heater tank assembly

earth terminal

terminal block

flow governor

solenoid valve

indicator lamp

microswitch

cam

inlet connection

switch assembly

cable entry

outlet connection

Heater shown is made for direct mains connection. Some types suited to storage-fed supplies.

Heater may be rated up to 8 kW. Need to ensure that electricity supply is adequate.

Flow governor will compensate for pressure variations.

Figure 3.5   Instantaneous electric shower heater

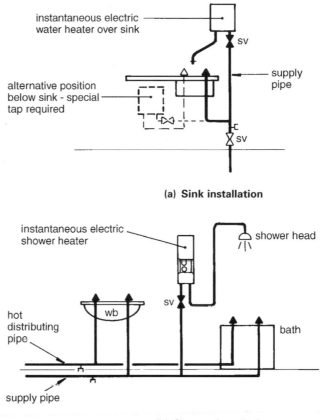

(a) Sink installation

(b) Shower installation

Where shower unit is fitted it will need a minimum head of 10.5 m.

Flexible shower outlet may be a contamination risk if nozzle can become submerged (see figure 3.51b).

Figure 3.6   **Typical uses for instantaneous electric water heaters**

3.3   Water-jacketed tube heaters

The water drawn off for use passes through a heat exchanger in a reservoir of primary hot water (see figure 3.7). The size of this reservoir and its heat input determine the volume and rate of flow of hot water that can be provided without an unacceptable temperature drop. The cold water feeds to the heater may be from the mains or from a storage cistern.

Water-jacketed tube heater is a form of instantaneous heater.

Primary circuit may be vented system or sealed system.

Heat exchanger warms secondary supply water as it passes through.

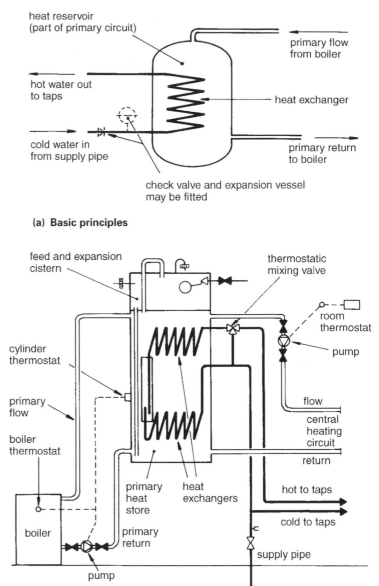

(a) Basic principles

(b) Use of water-jacketed tube heater

Note 1   Primary water from the boiler flows to the heat store as programmed by the cylinder thermostat. Hot water is pumped to the radiator heating circuits and returns to the heat store. Cooler water from the heat store then returns to be reheated in the boiler.

Note 2   Cold water under mains pressure from the supply pipe enters the lower heat exchanger to be partially heated. It then passes through the upper heat exchanger where it is fully heated before being distributed to the taps.

Drawing shows the 'Boilermate' system using a combination unit.

Figure 3.7   Water-jacketed tube heater

Combination boilers

Combination boilers, more commonly known as combi's, provide hot water on demand and full central heating all from one centralized heating source (see figures 3.8 and 3.9). Hot water is heated indirectly using the principle of the water-jacketed tube heater in its water-to-water heat exchanger.

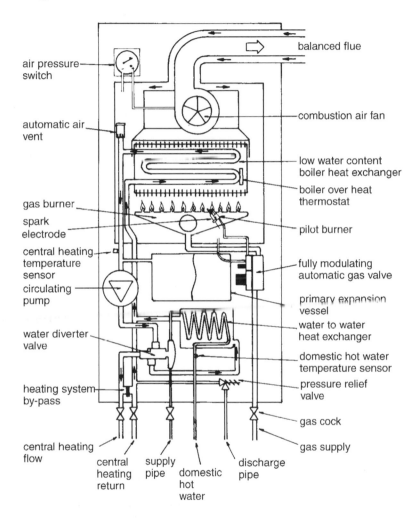

Figure 3.8   **Typical combination boiler**

BS 6700 does not discuss combination boilers although a large proportion of the heating and hot water market favours this type of system both for new and replacement work.

Features of the combination boiler include the following:

- It provides instant hot water on demand at a constant temperature.
- Water is only heated as and when it is needed.

System is relatively simple and cheap to install.

Heating circuit usually of the 'sealed' type.

No system controls or valves shown (see figure 3.7 (c)).

Temporary fill point must be disconnected from supply pipe after filling or topping up.

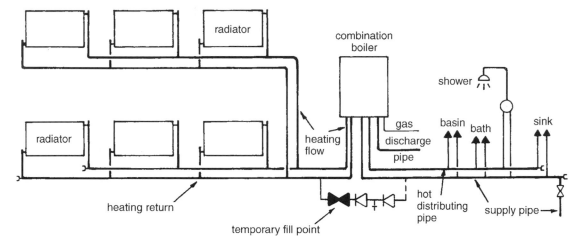

Figure 3.9   **System using combination boiler**

- Cold water storage is eliminated, saving on materials and installation costs, but there is no reserve water provision.
- Electronic modulating regulation provides for a wide range of heating loads.
- It provides constant flow rate of 8 l to 13.5 l per minute depending on boiler output rating and on the adequacy of the incoming water supply.
- The sealed system of the central heating primary circuit gives priority to hot water when needed.

3.4   Storage type water heaters and boiler heated systems

Domestic hot water supply installations of the storage type may be either vented or unvented. Figures 3.10 and 3.11 illustrate the main features of each.

System choice will be made in conjunction with choice of cold water system, and will depend on the need, or otherwise, to maintain a reserve of cold water to keep the system in operation at all times, or a preference to avoid any disadvantages of high-level cold water storage.

Vented hot water storage system

This is fed from a high-level cistern which provides the necessary pressure at outlets, accommodates expansion due to heated water, and is fitted with

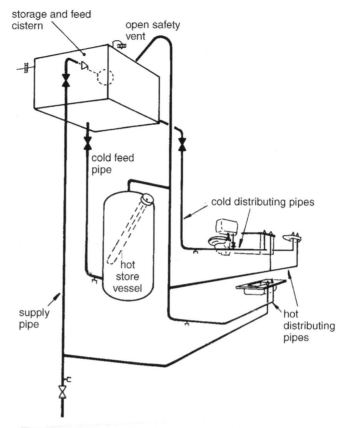

The open vent pipe will provide safety relief in the event of the system overheating.

Heated water will expand to the cold feed pipe.

The system shown is heated by immersion heater but could alternatively be heated by boiler.

Figure 3.10   **Vented hot water system**

an open safety vent pipe to permit the escape of air or steam, and to prevent explosion without the need for any mechanical device.

The vented hot water storage system provides:

- constant low pressure;
- reserve water supply;

but needs;

- protection against the entry of contaminants to cistern (a protected drinking water cistern that meets the requirements of Water Regulations will guard against this).

Unvented hot water storage system

This is usually fed direct from the supply pipe under mains pressure. It does not require the use of a feed cistern or open vent pipe, but relies on mechanical devices for the safe control of heat energy and hot water expansion.

Building Regulations Approved Document G3 requires that hot water shall not exceed 100°C. The Approved Document recommends that all

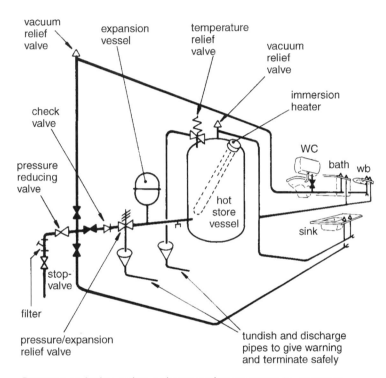

Pressure reducing valve reduces mains pressure to pressure suitable for operating system.

Pressure relief valve guards against excess pressure.

Expansion vessel accommodates expansion of water when heated.

Thermostat (not shown) controls temperature at normal level.

High temperature cut-out (not shown) protects against over-heating of water.

Temperature relief valve allows boiling water and steam to escape if thermostat and thermal cut-out should fail.

Tundish and discharge pipes take relief water to safe place.

(a)   Basic outline of system and components

Figure 3.11   Unvented hot water storage system

continued

Thermostat and high temperature cut-out fitted but not shown.

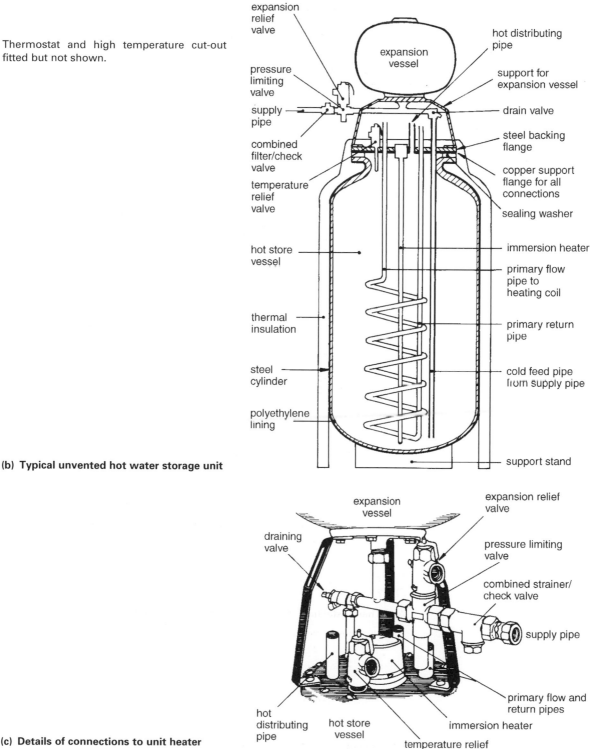

(b) **Typical unvented hot water storage unit**

(c) **Details of connections to unit heater**

Figure 3.11   Unvented hot water storage system continued

unvented hot water storage systems should be of the 'unit' or 'package' type, supplied complete with all safety devices, and should be fitted by a competent installer. The system should be approved by either:

(a) a member body of the European Organisation for Technical Approvals (EOTA), e.g. the British Board of Agrément (BBA); or
(b) a certificating body having NACCB accreditation (National Accreditation Council for Certification Bodies), e.g. complying with BS 7206; or
(c) independent assessment which clearly demonstrates that an equivalent level of verification and performance as in (a) and (b) is achieved.

A *unit system* (see figure 3.11b) has all safety devices and other operating devices fitted by the manufacturer at the factory, ready for site installation.

A *packaged system* (see figure 4.16) has all safety devices factory fitted. However, all other operating devices are supplied by the manufacturer in 'kit' form for site assembly.

Features of the unvented hot water storage system are as follows:

- needs continuity of mains supply pressure and flow, otherwise hot water cannot be guaranteed;
- eliminates the need for cold water storage and risk of frost damage;
- may require a larger supply pipe but eliminates some duplication of pipework;
- contains no reserve supply in case of supply failure;
- eliminates cistern refill noise;
- relies on mechanical safety devices for protection from explosion which need regular inspection and maintenance;
- gives better pressure at outlets, particularly at showers;
- allows quicker installation than vented system but involves more costly components.

Non-pressure or inlet-controlled water heaters

These are generally seen as single point heaters fitted either above the appliance with a swivel outlet spout, or under the appliance using special taps to control the flow before the heater inlet. They may be heated by either gas or electricity. See figure 3.12.

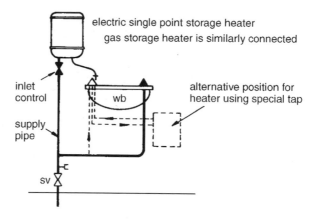

electric single point storage heater
gas storage heater is similarly connected

inlet
control

alternative position for
heater using special tap

wb

supply
pipe

sv

Expansion of heated water overflows through outlet spout.

Outlet must not be obstructed, nor must any connection be made to it.

Can be connected to wash basin or sink if special taps are used to control the inlet and leave the outlet unobstructed.

Figure 3.12   **Non-pressure or inlet-controlled water heater**

Pressure or outlet-controlled water heaters

These may be heated by either gas or electricity. Although these are called pressure type heaters, they are generally designed for supply from a feed cistern and are not usually suitable for operation under direct mains pressure. See figures 3.13 and 3.14. Care should be taken to ensure that the heater will withstand the pressures to which it is to be subjected. In smaller dwellings a capacity of 100 l to 150 l is considered sufficient except where off-peak electricity is used this figure can then be doubled.

Cold feed pipe provides for expansion of heated water.

Open safety vent terminates over feed cistern to provide safe route for boiling water and steam, should the system overheat.

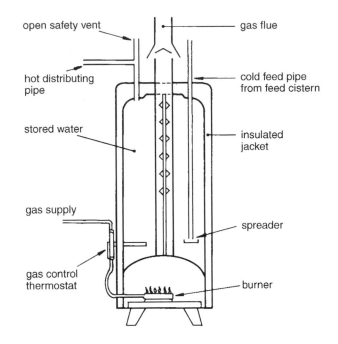

Figure 3.13   **Outlet-controlled gas water heater**

For dwellings, cistern must be 'protected'.

Heater capacity (minimum):
- for small dwellings, 100 l to 150 l,
- for off-peak electricity systems, 200 l.

Electricity systems are usually in the form of factory lagged and cased hot water cylinder with connections similar to those shown here for a gas installation.

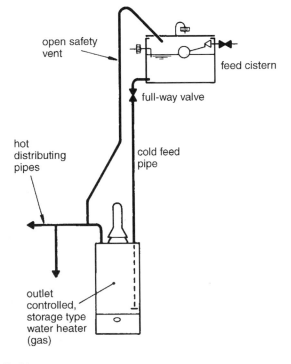

Figure 3.14   **System using outlet-controlled heater**

Cistern or combination type storage heaters

A combination type storage water heater incorporates a cold water cistern which should be located so that at the very least its base is not lower than the level of the highest connected hot water outlet, and is high enough to give adequate flows at outlets, see figures 3.15 and 3.16.

Direct type will often be used with an electric immersion heater as the sole source of heat.

Flow and return from heat exchanger may be connected to centralized boiler for main heat source, with immersion heater used as supplementary heat source.

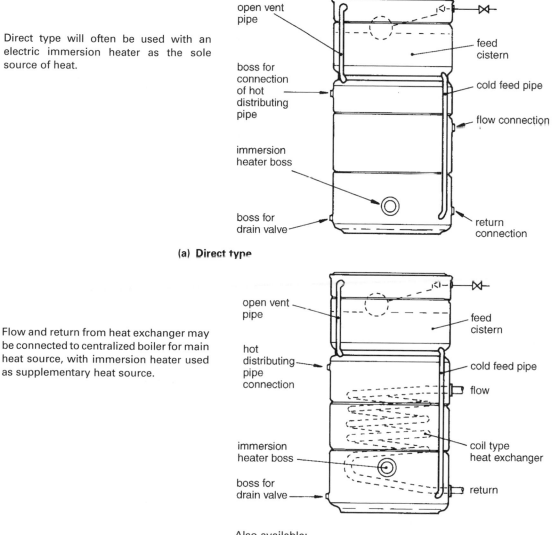

(a) **Direct type**

Also available:
- ○ factory lagged or purpose-made unit with lagging and metal outer casing;
- ○ single feed indirect type.

(b) **Double feed indirect type**

Figure 3.15   **Combination storage heaters**

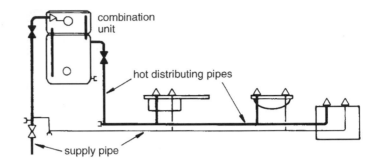

Unit can be directly heated by immersion heater or indirectly heated from boiler.

Advantages:
○ low installation costs;
○ useful for flats when space is limited.

Disadvantages
○ cold storage space limited;
○ low pressure at hot taps;
○ cannot be used for showers.

Figure 3.16   Typical combination unit installation

Electric immersion heater type storage heaters

Immersion heaters can be used as an independent heat source, or to provide supplementary heat to other centralized boiler systems, see figures 3.17–3.20 for examples of vented systems.

Systems using immersion heaters may be vented or unvented. Safety devices must be used as appropriate to the system.

Any immersion heater used should comply with BS 3456: Section 2.21; whilst electrical controls should comply with BS 3955.

Immersion heaters and controls should be located so they are readily accessible for insertion, removal and adjustment. Figure 3.18 gives alternative immersion heater positions.

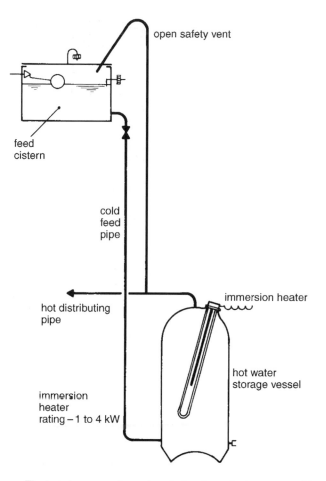

open safety vent

feed
cistern

cold
feed
pipe

hot distributing
pipe

immersion heater

immersion
heater
rating – 1 to 4 kW

hot water
storage vessel

The hot store vessel may be a hot water cylinder or a combination
unit, or a specially made, encased and lagged cylinder.

Figure 3.17   Vented storage system using immersion heater

(a) Single element – top entry

(b) Double element – top entry

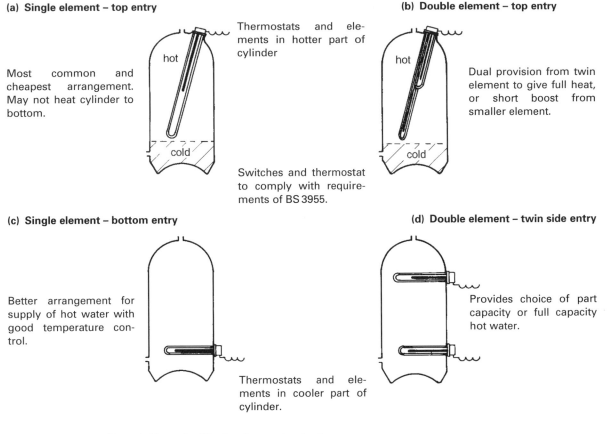

Most common and cheapest arrangement. May not heat cylinder to bottom.

Thermostats and elements in hotter part of cylinder

Dual provision from twin element to give full heat, or short boost from smaller element.

Switches and thermostat to comply with requirements of BS 3955.

(c) Single element – bottom entry

(d) Double element – twin side entry

Better arrangement for supply of hot water with good temperature control.

Provides choice of part capacity or full capacity hot water.

Thermostats and elements in cooler part of cylinder.

Figure 3.18   **Positioning immersion heaters**

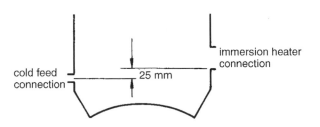

Immersion heater connection to be at least 25 mm above the centre line of the cold water inlet.

Adequate clearance needed around immersion heater for removal and replacement.

Figure 3.19   **Immersion heater position and cold feed connection**

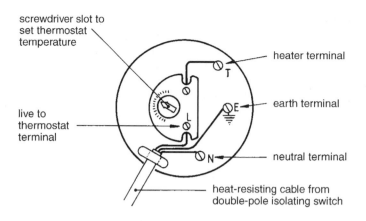

Range of temperature settings – 50°C to 82°C. Suggested setting 60°C.

Figure 3.20   **Top of immersion heater with cover removed**

Gas-fired circulators

These are essentially small gas-fired boilers. They may be independent or used in conjunction with a system using some other fuel, see figure 3.21.

In hard water areas use indirect system. Otherwise, direct or indirect system may be used.

Economy valve gives choice of whole or part cylinder heating.

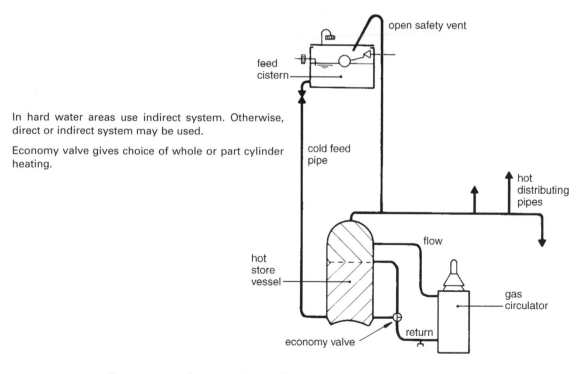

Figure 3.21   **System using gas-fired circulator**

Boiler-heated hot water systems

BS 6700 deals with independent water heating appliances (boilers) using gas, oil, solid fuel and electricity, and includes direct, indirect, vented and unvented systems.

Direct systems are those in which the boiler or heat source 'directly' heats the water that is drawn off. The principles of this system are shown in figure 3.22. The electric immersion heater system shown in figure 3.17 is also a type of direct system.

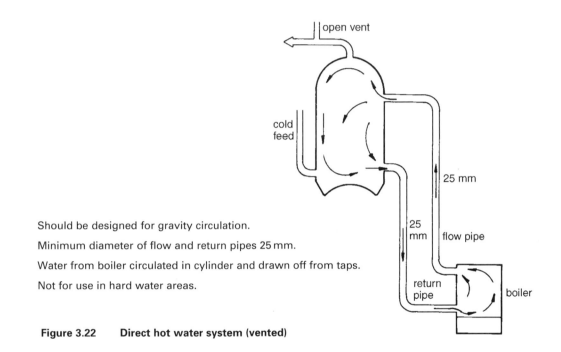

Should be designed for gravity circulation.

Minimum diameter of flow and return pipes 25 mm.

Water from boiler circulated in cylinder and drawn off from taps.

Not for use in hard water areas.

Figure 3.22   **Direct hot water system (vented)**

In **indirect systems** the primary water (heated by the boiler) is physically separated from the secondary water (water drawn off from the taps) by a heat exchanger in the form of a coil, or inner annular cylinder, see figure 3.23.

Where systems combine both hot water and central heating functions indirect systems are preferred to prevent excessive furring in hard water areas, to maintain a cleaner secondary hot water supply and to permit the use of corrosion inhibitors, thus increasing the life of components.

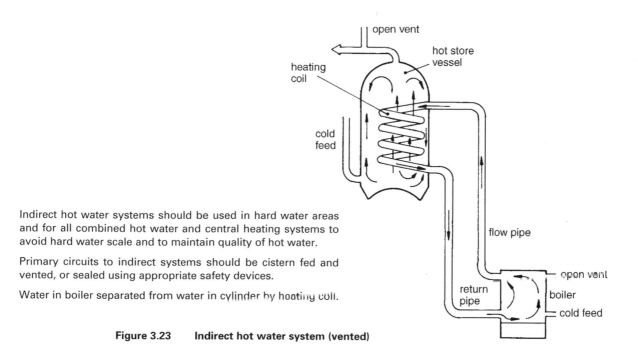

Indirect hot water systems should be used in hard water areas and for all combined hot water and central heating systems to avoid hard water scale and to maintain quality of hot water.

Primary circuits to indirect systems should be cistern fed and vented, or sealed using appropriate safety devices.

Water in boiler separated from water in cylinder by heating coil.

Figure 3.23 **Indirect hot water system (vented)**

Gravity flow hot water circulation

Circulation in hot water systems may be attained by using natural convection currents (gravity circulation) or, alternatively, by the use of a circulation pump to 'force' the water around the system. Hence the terms 'forced circulation' and 'pumped circulation'. Gravity circulation is dealt with initially.

In gravity circuits, good circulation depends on system height and the difference in densities between the hot flow and cooler return to provide the motive force to overcome frictional resistances in pipes and fittings and to provide effective hot water circulation.

Figure 3.24 shows examples of good and poor circulation in gravity primary circuits and table 3.1 relates the density of water to various temperatures.

Table 3.1 **Examples of water temperature and density**

Temperature °C	Density kg/m^2
4	1000
40	992
50	988
60	983
70	978
82	974

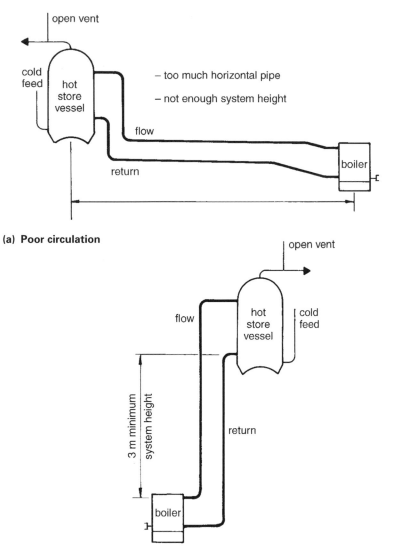

(a) Poor circulation

In gravity circuits good circulation depends on system height and pressure differentials between hot flow and cooler return to overcome resistances in pipes and bends.

(b) Good circulation

Figure 3.24   Circulation in primary circuits (gravity)

Example of calculation of circulating head in gravity circuits

The circulating head (CP) can be calculated by the formula:

$$CP = (Dr - Df)\, 9.81 \times \text{circulating height (m)}$$

where

Dr = density of return pipe (kg/m^3);
Df = density of flow pipe (kg/m^3);
9.81 = value of gravitational force in Newtons (N)

So taking figure 3.24(b) as an example and assuming a flow and return temperature of 82°C and 60°C, respectively:

$$CP = (Dr - Df)\, 9.81 \times \text{circulating height (3 m)}$$
$$= (983 \times 974) \times 9.81 \times 3$$
$$= 264.87 \, \text{N/m}^2 \text{ or } 264.87 \, \text{Pa}$$

3.5   Primary circuits

Primary circuits are used for circulation of hot water between the boiler and the hot store vessel and include any radiator circuits (see figures 3.25 and 3.26). They may be vented or unvented. Circulation may be gravity or pumped, direct or indirect.

Direct systems are rarely used in modern systems because of the problems of hard water scale and corrosion in primary circuits due to the constant changing of the water. Corrosion may also lead to discoloration of water drawn off from taps.

Vented primary circuits are required to have two important pipe connections in addition to those for the primary flows and return circulating pipes.

(1)   The first is a permanent 'safety' vent route from the hottest, topmost part of the boiler to a point terminating above and into the feed and expansion cistern. This will provide for the immediate release of boiling water and steam should the boiler overheat.

(2)   The second connection is the cold feed pipe which provides a route for filling and replacing primary system water, and also provides a route for the expansion of water when heated, the excess being accommodated in the feed and expansion cistern.

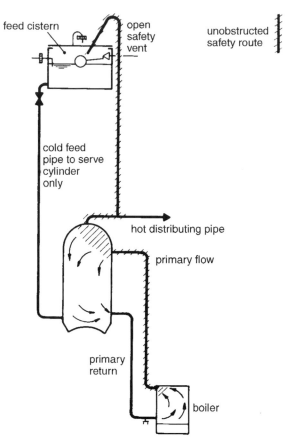

feed cistern

open
safety
vent

unobstructed
safety route

cold feed
pipe to serve
cylinder
only

hot distributing pipe

primary flow

primary
return

boiler

All pipes to be laid to falls to avoid air locks and facilitate draining.

Vent route from top of boiler through cylinder to open vent not to be valved or otherwise closed off.

Minimum sizes for primary circuits to hot store vessels:
- ○ 25 mm for solid fuel gravity systems
- ○ 19 mm for small bore systems

Corrosion inhibitors not to be used.

Figure 3.25   **Direct system of hot water (vented)**

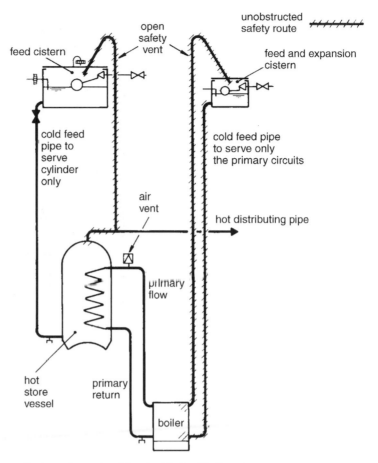

System shown is of the double feed indirect type.

Open vent and cold feed pipes to primary circuit should *not* have valve, or be otherwise closed off.

The open vent and cold feed pipes may be connected to the primary flow and return pipes respectively.

Where vent pipe is not connected to highest point in primary circuit, an air release valve should be installed.

Figure 3.26   **Indirect system of hot water (vented)**

Hot water systems using the single feed indirect cylinder

The single feed indirect cylinder permits cost savings on materials and installation because there is no need for a separate open vent and cold feed for the primary circuits, see figures 3.27 and 3.28.

They are, however, limited in use as the volume of the cylinder is directly related to the volume of the water in the primary circuit (including radiators). If the system is oversized or becomes overheated they might lose their water seals (air locks) and revert to direct circulation.

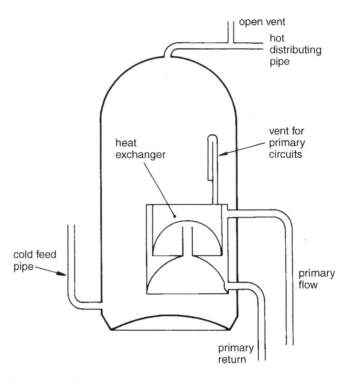

open vent

hot distributing pipe

vent for primary circuits

heat exchanger

cold feed pipe

primary flow

primary return

Expansion of heated water is taken up in inner heat exchanger, which must be matched by system size. Oversizing of system will result in loss of seals and mixing of primary and secondary waters. Working principles shown in figure 3.41.

Reference should be made to manufacturers regarding suitability for use with boilers and radiators.

Figure 3.27   Single feed indirect cylinder to BS 1566

Single feed indirect cylinders must be vented.

BS 6700 gives the following recommendations for the use of single feed indirect cylinders:

(a) Cylinders should conform to BS 1566: Part 2 and be installed as per manufacturers' instructions.
(b) Where the primary circuit is pumped, the static head must not exceed the pump head.
(c) Corrosion inhibitors or other additives must not be added to primary circuits.
(d) Boiler and radiator manufacturers' recommendations must be followed.

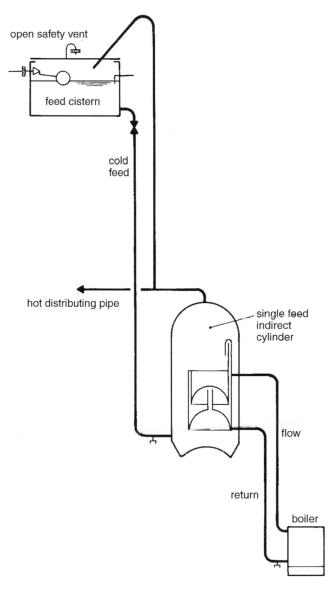

open safety vent

feed cistern

cold
feed

hot distributing pipe

single feed
indirect
cylinder

flow

return

boiler

Installation similar to direct system.

When system is pumped, static head should exceed pump head.

No corrosion inhibitor to be added.

Figure 3.28   System using single feed indirect cylinder

Classes and grades

Single feed indirect cylinders are pressure graded and should be used within the pressure ranges shown later in Table 3.5. Their primary heat exchangers may be one of two grades based on the maximum permissible capacity of the primary circuit including pipes, boiler and any radiators installed.

- Class 110 – for systems having a primary capacity of not more than 110 l.
- Class 180 – for systems having a primary capacity of not more than 180 l.

Combined systems of hot water and central heating

It is usual practice in domestic and many non-domestic premises to combine both hot water and central heating functions in one system using a centralized heat source, e.g. boiler, with in some cases a supplementary heat source, such as a gas circulator or electric immersion heater for summer use.

The primary circuits to these systems will invariably incorporate the use of a circulating pump to distribute some or all of the heat produced by the boiler. Circulation methods for these systems may include:

- gravity hot water and pumped central heating;
- fully pumped hot water and central heating.

These systems are illustrated in figures 3.29–3.33.

There is an increasing tendency to use the fully pumped system which has the advantage of a smaller circulating pipework leading to lower costs and quicker installation times, and a quicker, more efficient system heat-up.

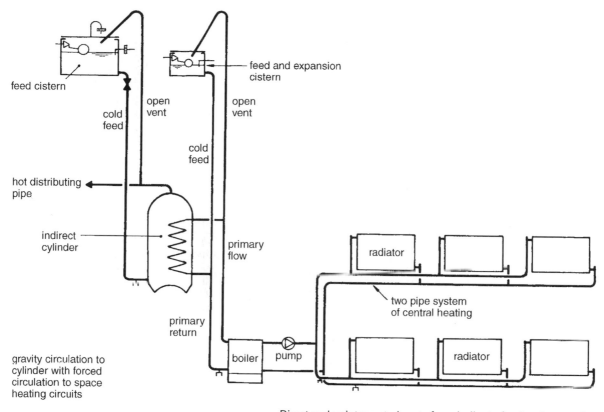

feed cistern

feed and expansion cistern

open vent

open vent

cold feed

cold feed

hot distributing pipe

indirect cylinder

primary flow

primary return

gravity circulation to cylinder with forced circulation to space heating circuits

radiator

radiator

two pipe system of central heating

boiler

pump

Direct and uninterrupted route from boiler to feed and expansion cistern for cold feed and open vent pipes – also through open vent from cylinder.

Figure 3.29   Combined hot water and central heating system with vented primary circuits

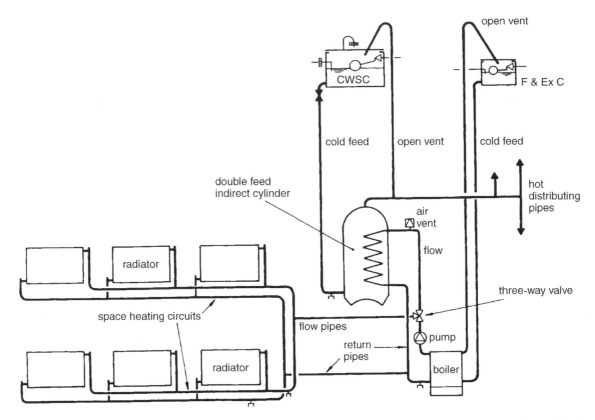

open vent

CWSC

F & Ex C

cold feed   open vent   cold feed

double feed
indirect cylinder

hot
distributing
pipes

air
vent

flow

radiator

three-way valve

space heating circuits

flow pipes

return
pipes

pump

radiator

boiler

Cold feed and vent pipes uninterrupted between boiler and feed
and expansion cistern.

Figure 3.30   Fully pumped system of hot water and central heating (vented)

Sealed central heating systems with vented hot water

Sealed (unvented) primary circuits are often associated with the combination boiler with which they are commonly used. They can, however, be used with both vented and unvented secondary hot water systems as an alternative to the vented primary circuits previously discussed.

Sealed systems require the use of an expansion vessel to accommodate the increased volume of water when the system is heated through temperatures between 10°C and 100°C.

Sealed systems also require the use of a number of mechanical safety devices which are indicated in figure 3.31.

Primary heaters (coil or annular cylinder) should be suitable for operation at pressures of 0.35 bar above the pressure relief valve setting.

Indirect cylinder suitable for pressures of up to 0.35 bar in excess of relief valve setting.

Requires the use of mechanical devices for safety in the control of temperature and expansion of heated water.

Sealed system may be used with gravity primaries to cylinder.

Detail to figure 3.31 showing operating principle of expansion vessel

Expansion vessel must accommodate expansion of the total volume of water in the whole system including space heating circuits (approximately 4%).

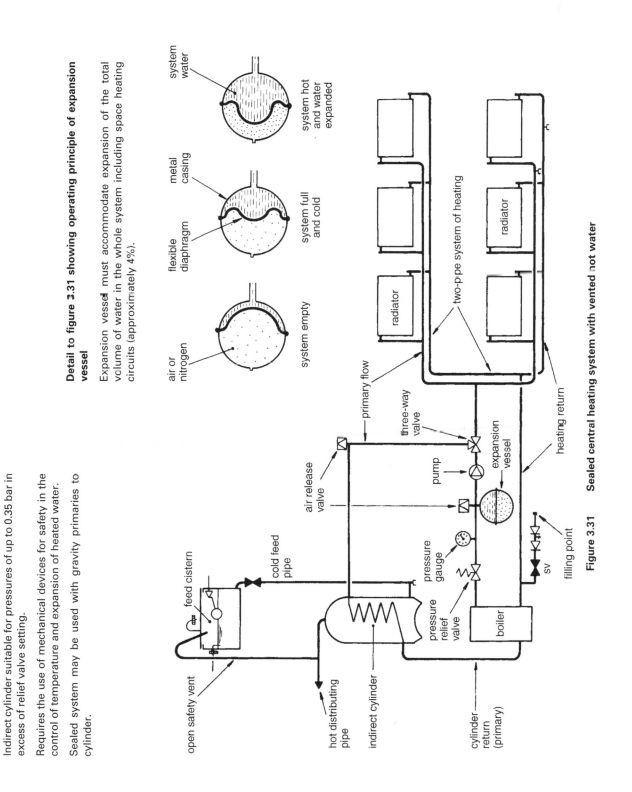

Figure 3.31   Sealed central heating system with vented hot water

Combined hot water and central heating with common return

Where a system has gravity flow to the hot store vessel with pumped central heating circuits, BS 6700 recommends that both flow and return connections from each circuit be connected separately to the boiler.

Unfortunately, not all boilers have enough tappings to accommodate separate connections. In these cases the return pipes may be combined by use of an injector tee as shown in figures 3.32 and 3.33. Otherwise the solution is to use a fully pumped system as in figure 3.30.

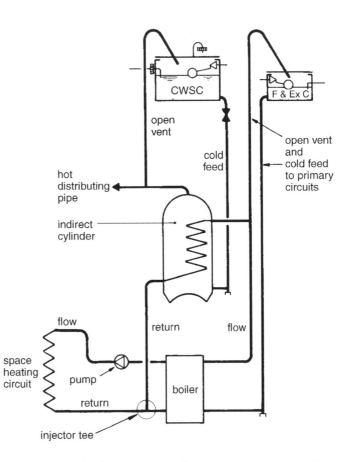

Where return circuits are connected via a common return to the boiler and only the heating circuit is pumped, an injector tee should be fitted near to the boiler connection.

Figure 3.32   **Combined hot water and central heating with common return**

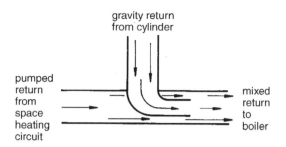

gravity return
from cylinder

pumped
return
from
space
heating
circuit

mixed
return
to
boiler

Figure 3.33   **Detail of injector tee**

Supplementary water heating and independent summer water heating

It is common practice for supplementary water heating and independent summer water heating to be provided by use of electric immersion heaters in the storage vessel or by gas circulators. Supplementary water heating provided from solar energy or heat pumps is growing in popularity.

Where supplementary electric immersion heating is to be used in conjunction with a boiler, the storage vessel should be positioned at least 1 m above the boiler to prevent circulation of hot water from the storage vessel to the boiler when only the immersion heater is in use.

Solar heating

Solar energy heating can be useful to augment a conventional domestic hot water system (see figure 3.34) but in Britain, except on hot sunny days, cannot be considered sufficient to heat hot water to draw-off temperatures.

Many different designs of solar systems for heating water are possible, from simple direct feed gravity systems to more complex pumped circuits using two indirect storage cylinders or one cylinder with two indirect heating coils. Therefore, apart from indicating the possibilities of their use, they are beyond the scope of this book.

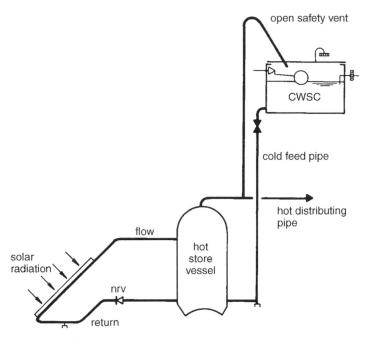

Drawing shows basic principles only.

Figure 3.34   **Solar heating system for hot water**

Heat pumps

Heat pumps extracting energy from the ground, water, or air at ambient temperatures, can be used to preheat conventional hot water systems, to augment existing systems, or to supply full hot water and central heating requirements, see figure 3.35.

Until such time as a British Standard is available giving recommendations for their use, guidance should be sought from heat pump manufacturers, or the Heat Pump Association, 161 Drury Lane, London WC2B 5QG.

3.6   Secondary hot water distributing systems

Secondary hot water distributing systems include any hot store vessel, pipes and components used to store the hot water and convey it to the point of use.

A secondary hot water distribution system may be one of the following types:

(1)   gravity fed (vented) from a cold water feed cistern;
(2)   gravity fed (unvented) from a water storage cistern;
(3)   directly supplied under pressure from mains, through an instantaneous water heater or a water jacketed tube heater;

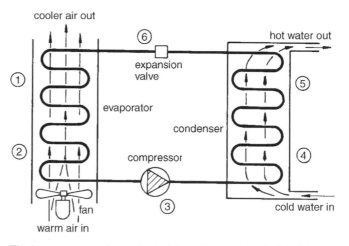

The heat pump works on the 'refrigeration cycle' principle (air to water)

(1) Refrigerant liquid forced under pressure into evaporator.
(2) Heat energy from air absorbed by refrigerant as it is vaporized from liquid to gas.
(3) Compressor 'squeezes' gas which becomes hotter and more compact.
(4) The hotter compacted gas is pumped under pressure into the condenser.
(5) In the condenser the gas becomes liquid and throws off heat which is absorbed into the hot water.
(6) Condensed refrigerant liquid passes through expansion valve, to lower its pressure, before flowing through to evaporator to begin the cycle again.

Figure 3.35   **The heat pump – principle**

(4)   unvented storage type, directly supplied under pressure from mains.

Connections to hot water storage vessels should be arranged so that:

(a) The cold feed pipe is connected near the bottom, below any primary return connection or central heating element, thus ensuring that only the hot water is drawn off.
(b) The hot distributing pipe is connected to the top of the cylinder above any primary flow connection or heating element. This will minimize mixing of hot and cold water during draw-off and refilling, and minimize cold shock to the heating element.

Dead legs and secondary circulation

To promote maximum economy of energy and water the hot water distributing system should be designed so that hot water appears quickly at draw-off taps when they are opened. The length of pipe measured from the tap to the water heater or hot water storage vessel should be as short as

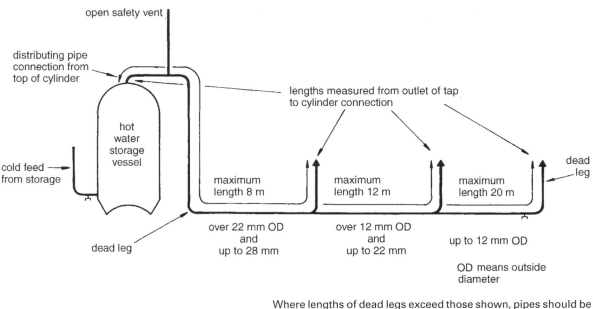

Where lengths of dead legs exceed those shown, pipes should be insulated. In general, however, it is better for all pipes to be insulated.

Note 'Dead leg' means the length of distributing pipe without secondary circulation.

Figure 3.36   **Maximum length of dead legs without insulation**

possible and should not exceed the lengths shown in figure 3.36. Where these lengths exceed the values given in table 3.2 the pipe should be insulated to at least the standards given in tables 3.3 and 3.4.

Building Regulations require that pipes should be insulated with a material that has a thermal conductivity value of 0.045 W/m.K or less and a thickness equal to the pipe diameter, up to a maximum of 40 mm.

Secondary circulation minimizes delays in obtaining hot water from taps and reduces waste of water during any delay. Secondary circulation should be considered when short dead legs are impractical and the circuit should be well insulated to reduce the inevitable heat losses from pipe runs. A diagrammatic arrangement of secondary circulation is shown in figure 3.37.

In systems where it is not possible to attain gravity circulation, a non-corroding circulating pump should be installed to ensure that water within the secondary circuit remains hot. The pump should be located on the return pipe close to the cylinder.

Secondary returns should connect to the top quarter of the hot store vessel to take best advantage of the hottest water.

To economize on energy and prevent circulation during off-peak periods, a night valve should be fitted to the return near to the cylinder connection. This may be manually operated, but preferably an electrically operated time control valve will be used.

Table 3.2 **Maximum lengths of uninsulated distributing pipes**

Outside diameter of distributing pipe mm	Maximum length m
Not exceeding 12	20
Exceeding 12 but not exceeding 22	12
Exceeding 22 but not exceeding 28	8
Over 28	3

Table 3.3 **Economic thickness of insulation for domestic hot water systems in heated areas**

Outside diameter of copper pipe (in mm)	Water temperature of +60°C with ambient still air temperature of +20°C				
	Thermal conductivity at +40°C (in W/(m.K))				
	0.025	0.030	0.035	0.040	0.045
	Thickness of insulation (in mm)				
10	13	14	14	14	15
12	13	14	14	15	16
15	13	14	14	16	17
22	14	15	16	17	18
28	14	15	16	18	19
35	15	17	17	19	19
42	15	17	18	19	20
54	16	18	19	20	21
Flat surfaces	20	22	24	24	25

Table 3.4 **Economic thickness of insulation for domestic hot water systems in unheated areas**

Outside diameter of copper pipe (in mm)	Water temperature of +60°C with ambient still air temperature of −1°C				
	Thermal conductivity at +40°C (in W/(m.K))				
	0.025	0.030	0.035	0.040	0.045
	Thickness of insulation (in mm)				
10	14	15	16	17	18
12	15	16	17	18	19
15	15	17	17	19	19
22	16	18	20	20	21
28	17	19	20	21	30
35	18	20	21	22	31
42	19	20	22	23	32
54	20	21	23	33	34
Flat surfaces	23	25	25	29	31

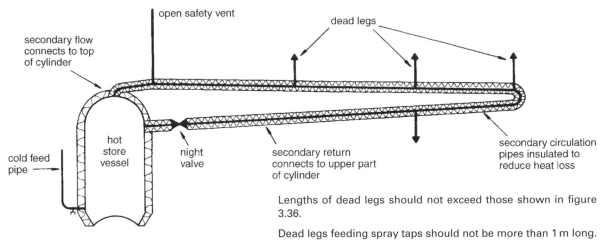

open safety vent

dead legs

secondary flow connects to top of cylinder

cold feed pipe

hot store vessel

night valve

secondary return connects to upper part of cylinder

secondary circulation pipes insulated to reduce heat loss

Lengths of dead legs should not exceed those shown in figure 3.36.

Dead legs feeding spray taps should not be more than 1 m long.

Figure 3.37   Distributing system with secondary circulation

3.7   Components for hot water systems

Cold feed pipes

A cold feed pipe supplies hot water apparatus with cold water from storage (see figure 3.38). In vented systems of secondary hot water, the cold feed water is supplied from a feed cistern at high level and should be connected near the bottom of the hot store vessel. A servicing valve that will not inhibit or restrict the flow should be fitted to the cold feed pipe as near as is practical to the feed cistern.

The cold feed pipe should not be used to supply any fitting other than the hot store vessel. In a direct system, the cold feed pipe should connect directly to the boiler and not to the return pipe.

In a vented indirect system, other than one using a single feed indirect hot water cylinder, the primary circuit must be supplied through a separate cold feed pipe from a separate feed and expansion cistern. The cold feed pipe should be connected to the lowest point in the primary circuit near to the boiler (see figure 3.30) and preferably to the boiler.

Alternatively, in systems having gravity circulation to the hot store vessel, it is permissible to connect to the gravity return as shown in figure 3.29. Where feed and expansion cisterns have a capacity of 18 l or more a servicing valve should be fitted on the cold feed pipe near to its connection to the feed and expansion cistern.

On smaller systems this valve is not necessary.

The open vent pipe

The open safety vent, as it is often called, is connected to the top of the hot storage vessel and rises to terminate above the cold feed cistern. Recom-

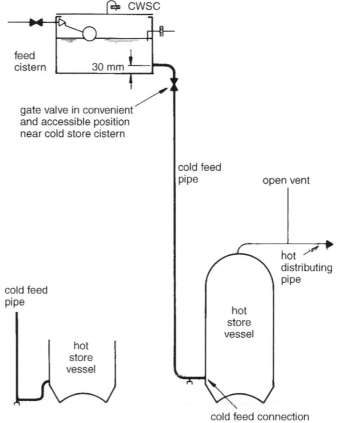

CWSC

feed
cistern

30 mm

gate valve in convenient
and accessible position
near cold store cistern

cold feed
pipe

open vent

hot
distributing
pipe

cold feed
pipe

hot
store
vessel

hot
store
vessel

cold feed connection
near bottom of cylinder

The dipped entry connection reduces the risk of hot water circulation in the cold feed pipe.

Cold feed must serve hot water apparatus only and no other
appliance.

The cold feed pipe must *not* be connected to the boiler on direct
systems.

No valve to be fitted to cold feed pipe on primary circuits.

Figure 3.38   Cold feed pipes

mendations for the installation of the open safety vent are given in figure
3.39.

Vented primary circuits are similarly treated. The vent must terminate
over a feed and expansion cistern. Due allowance should be made for any
head produced by a circulation pump (if fitted) in order to prevent either
continuous discharge from the vent, or air entrainment into the system.

The lower connection of the vent may be made to the highest point in
the primary flow pipe, or preferably to the top of the boiler.

Vent connection offset to reduce parasitic (one pipe) circulation and loss of heat in vent pipe. The minimum offset recommended is 450 mm.

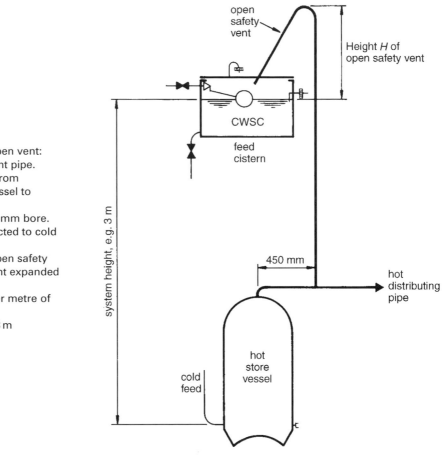

Rules for installation of the open vent:
○ No valve to be fitted to vent pipe.
○ Pipe to rise continuously from connection at hot store vessel to highest point over cistern.
○ Vent pipe not less than 19 mm bore.
○ Vent pipe *not* to be connected to cold feed.
○ Formula for height *H* of open safety vent over cistern to prevent expanded water from overflowing:
   $H = 150\,\text{mm} + 40\,\text{mm}$ per metre of system height
   For a system height of 3 m
   $H = 150 + (3 \times 40)$
      $= 270\,\text{mm}$

Note The formula shown above is for gravity systems. The height of the vent over the cistern in pumped systems should be related to the pump head.

Figure 3.39   The open safety vent

It should be noted that the open vent and cold feed pipes serve separate functions and should *not* be connected together.

Cylinders supplied from common feed cisterns

Common cold feed pipes are permissible, provided the installation is as shown in figure 3.40 and where individual separate systems are not practicable. This is an arrangement commonly used in flats under single ownership, i.e. local authority or housing association developments. It can

Separate vents are required to prevent hot water circulation from one cylinder to another.

Anti-siphon pipe is shown as an alternative backflow prevention device and is not needed if mechanical devices are used at each floor level.

Pipes to be aligned so as to avoid air locks and facilitate filling and draining.

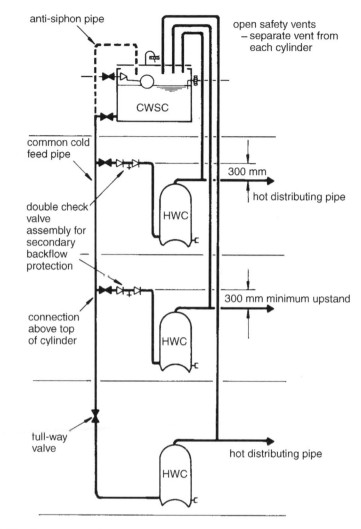

Figure 3.40   Common cold feed arrangement

lead to problems where flats become privately owned and where pipes and cisterns are a shared responsibility.

Hot water storage vessels

Hot water storage vessels are manufactured in a number of varieties and styles. Except in special cases, only cylinders and tanks to a relevant British Standard should be considered, preferably those with works-applied external insulation.

Building Regulations (Approved Document L) require that hot store vessels be efficiently heated and provide water at a suitable temperature for the user. Boiler-heated vessels should be fitted with an effective heat exchanger for the transfer of heat from the primary circuit to the secondary hot water. They should be fitted with a thermostatic control to shut off the heat when the recommended cylinder temperature is reached, and a time switch to shut off the heat supply when there is no demand.

It is important that the correct grade of cylinder or tank is fitted, having adequate wall thickness to suit the internal pressure to which the vessel will be subjected (see table 3.5).

Table 3.5   Grades and pressure rating for hot store vessels

Type and BS number	Grade	Max. head	Grade	Max. head	Grade	Max. head	Grade	Max. head
Galvanized steel cylinders and tanks:								
tank, direct to BS 417: Part 2	A	4.5 m	3.0 m	18 m	n/a	n/a	n/a	n/a
cylinder, direct to BS 417: Part 2	A	30 m	B	18 m	C	9 m	n/a	n/a
cylinder, indirect to BS 1565	1	25 m	2	15 m	3	10 m	n/a	n/a
Copper cylinders:								
direct to BS 699	1	25 m	2	15 m	3	10 m	4	6 m
double feed indirect to BS 1566: Part 1	1	25 m	2	15 m	3	10 m	4	6 m
single feed indirect to BS 1566: Part 2	n/a	n/a	2	15 m	3	10 m	4	6 m

n/a = not applicable.

Connection bosses to cylinders can be fitted to suit individual requirements, but 'standard' connections may include those shown in the various diagrams in this book.

To reduce the risk of corrosion to the cylinder or tank, a protector rod (sacrificial anode) may be fitted by the manufacturer. The need for this will depend on the type of water supplied and usually applies to deep well waters. The water supplier's advice should be sought before deciding on whether a protector rod is to be used.

The working principles of the single feed indirect type hot water cylinder are shown in figure 3.41.

Hot water storage vessels should be arranged so that the hotter water at the top floats on the relatively cold water, and hot water can be drawn off even though a substantial quantity of cold feed water may have recently flowed into the vessel.

Cylinders and tanks should be installed with the long side vertical to assist effective stratification or 'layering' of hot and cold water (see figure 3.42). The ratio of height to width or diameter should not be less than 2:1. An inlet baffle should be fitted, preferably near the cold inflow pipe, to spread the incoming cold water.

Hot water storage capacities should be related to likely consumption and the recovery rate of the hot water storage vessel.

For domestic use the temperature of stored hot water should not exceed 65°C and must not reach 100°C at any time. A temperature of 60°C should be adequate to meet all normal domestic requirements.

In other cases, such as in larger kitchens and in laundries, water temperatures may need to be higher. However, high temperatures should be avoided wherever possible because of furring in hard water areas and the danger of scalding users.

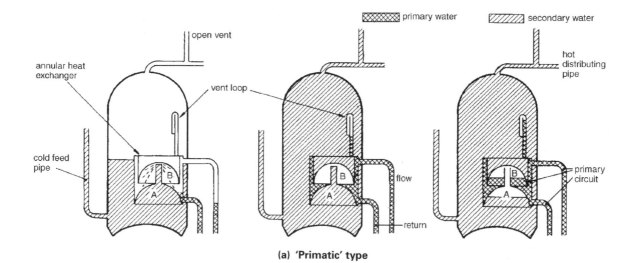

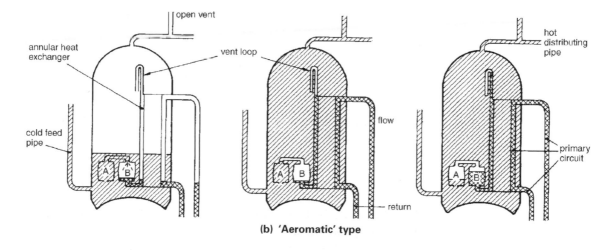

System filling Water from cold feed enters cylinder and from there into inner tank A. It then passes through inner tank B before finally filling the annular heat exchanger and the primary circuits. Air escapes from the vent loop.

System full Inner tank A is full. Inner tank B retains an air pocket. Annular heat exchanger is full and vent loop retains an air pocket. The air pockets prevent mixing of primary and secondary waters

System heating up Hot water expands in primary circuit and annular heat exchanger. The expanded water pushes the air pocket from tank B to be retained in tank A and the air pocket in the vent loop is moved but also retained.

Figure 3.41   **Single feed indirect cylinders – working principles**

In normal circumstances a temperature of 60°C will minimize scale formation and keep temperatures to a safe level for use by the young and the elderly whose skins are particularly sensitive to heat. See table 3.6 for domestic hot water requirements.

The capacity of the hot store vessel will depend on the likely con-

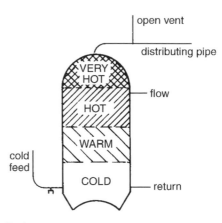

(a) Vertical cylinder

Vertical cylinder preferred to give better stratification and ensure that the hottest water is drawn off.

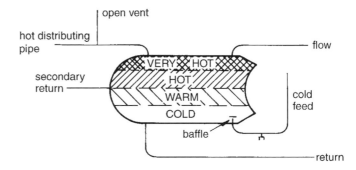

(b) Horizontal cylinder

Baffle will help prevent water from cold feed mixing with hot water at top of cylinder.

Hot water layer is very shallow and when drawn off may allow water layers to mix.

Figure 3.42   Cylinders showing stratification

Table 3.6   Hot water supply requirements

Requirement (dwellings)	Quantity l	Temperature °C	Flow rate l/s
Stored water per dwelling	135	60	n/a
Hot water used per person per day	35 to 45	Various	n/a
Bath per use	100	40	0.05 per tap
Shower per use	25	32 to 40	0.01 mixed
Wash basin per use	4.5	40 to 60	0.15 per tap
Sink per use	18	60	0.2 per tap

sumption and recovery rate of the vessel. Calculation of hot water storage capacity is dealt with later in section 5.4. In dwellings a quantity of 3.5 l to 4.5 l per person is recommended subject to a minimum storage capacity of 100 l for solid fuel boiler systems, and 200 l for systems using off-peak electricity. However, for speculative housing where occupancy is unknown, a storage capacity of 115 l is usual.

Insulation of hot water cylinders and tanks

In order to minimize energy consumption all hot water cylinders and tanks should be insulated (Building Regulations Approved Document L). For newly installed cylinders the best insulating medium is that applied as part of the cylinder manufacturing process. This will also provide some protection to the cylinder during transit to the site and during installation. For existing cylinders an insulating jacket to BS 5615 may be used, and fitted to ensure there are no exposed parts of the cylinder wall. Fixing bands should not be overtightened as this will reduce insulation efficiency.

Cisterns for supplying cold water to hot water apparatus

Feed cisterns used to supply the hot water system should conform to the requirements for cold water storage cisterns outlined in section 2.3. The feed cistern should be fitted at a height that will ensure satisfactory flow from all hot taps, and have a capacity at least equal to that of the hot store vessel that it feeds. A larger capacity will be required for combined storage and feed cisterns that also supply water to cold taps. Clause 2.3.9.1 of BS 6700 recommends at least 230 l minimum capacity for storage and feed cisterns, but this is inconsistent with other parts of the Standard which recommend 100 l only.

Feed and expansion cisterns to primary circuits (see figure 3.43) should comply with the requirements of BS 417 and BS 4213 and with the requirements of Water Regulations for cold water storage cisterns. The cistern should be situated so that its water level is below that of the feed cistern at all times (see figure 3.43b).

Overflow and warning pipes from feed cisterns and feed and expansion cisterns should run separately and should terminate outside the building in a prominent position.

Boilers

All boilers, firing equipment, and components of a combined boiler and cylinder unit should comply with appropriate British Standards wherever possible.

General provisions relating to boilers are as follows:

(1) Domestic boilers should be powerful enough to heat stored water to 65°C in 2.5 hours or less, including any towel rails, airing coils and secondary circulating pipes.

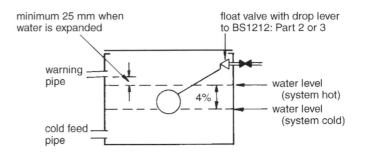

(a) **Space for expansion water**

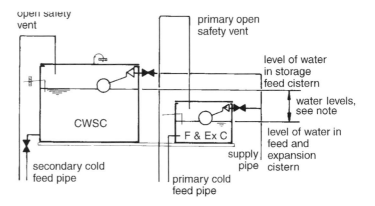

(b) **Water level in feed and expansion cistern**

Space must be allowed to accommodate expansion equal to 4% of water in circuit.

Cistern and float valve to resist temperature of 100°C.

Figure 3.43   **Feed and expansion cistern**

(2) Boilers and rooms containing boilers must be adequately ventilated to provide:
   (a) sufficient air for combustion;
   (b) enough air to remove the products of combustion and to discharge them safely and properly without danger to occupants of the building (see Building Regulations and the Gas Safety (Installation and Use) Regulations).
(3) Boilers and flues must be installed so they will not cause excess heat or fire in the building (see Building Regulations).
(4) Adequate space for access (see figure 3.44) is needed for:
   (a) boiler maintenance, removal of burners, pumps and pipe connections, and its eventual replacement;
   (b) stoking and cleaning, i.e. 1.25 times the back to front dimension of the boiler, subject to a minimum of 1 m.
(5) Precautions should be taken to ensure that combustible construction materials are not placed near boilers (see Building Regulations).

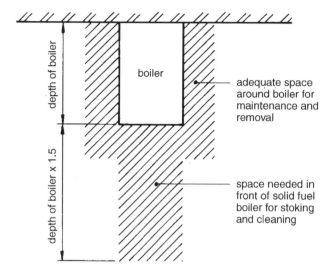

depth of boiler

depth of boiler × 1.5

boiler

adequate space
around boiler for
maintenance and
removal

space needed in
front of solid fuel
boiler for stoking
and cleaning

Figure 3.44   **Space requirements for access to boilers**

(6)   Fuel stores should be at a safe distance from the boiler (refer to fuel
suppliers' recommendations).

(7)   Precautions should be taken to prevent explosion should water temperatures exceed 100°C. An explanation of these precautions follows
in chapter 4.

(8)   In all cases, boiler manufacturers' recommendations should be followed for correct installation, commissioning and maintenance.

(9)   Solid fuel boilers produce heat energy for a long time after being
stoked and cannot readily be shut down. Sufficient allowance should
be made for the safe dissipation of that heat within the hot water or
heating system.

Boilers for use with sealed primaries and unvented systems require
specific controls and must therefore be selected only from those suitable
specifically for these purposes.

Gas-fired boilers should be appropriate for the gases with which they are
to be used.

Oil-fired boilers should be selected from the Domestic Oil Burning
Equipment Testing Association (DOBETA) list of tested and approved
domestic oil-burning appliances and installers. Typical boilers are shown in
figure 3.45.

Flues for boilers and water heaters

BS 6700 does not give details of requirements for flues and reference
should be made to various other codes and in particular to the Gas Safety
(Installation and Use) Regulations and to the Building Regulations. In
general terms the following three points should be considered:

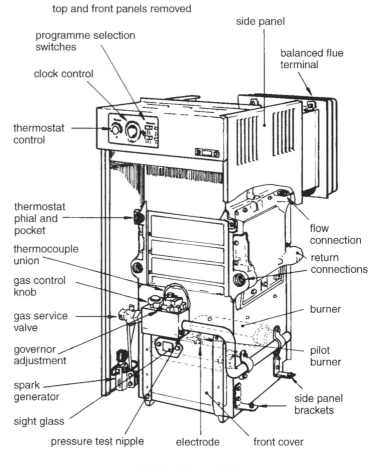

top and front panels removed

programme selection switches

clock control

thermostat control

side panel

balanced flue terminal

thermostat phial and pocket

thermocouple union

gas control knob

gas service valve

governor adjustment

spark generator

sight glass

flow connection

return connections

burner

pilot burner

side panel brackets

pressure test nipple   electrode   front cover

(a) Typical gas boiler

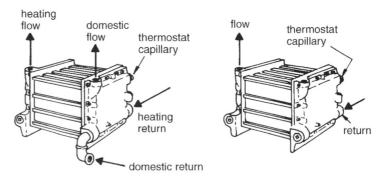

heating flow

domestic flow

thermostat capillary

heating return

domestic return

flow

thermostat capillary

return

(b) Boiler connections

Figure 3.45   Boilers

continued

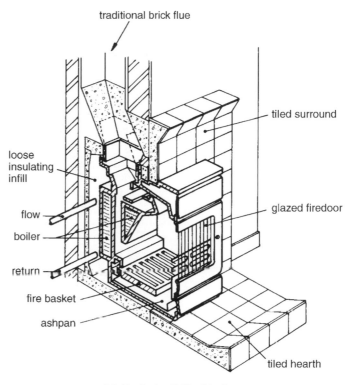

traditional brick flue

tiled surround

loose
insulating
infill

flow

boiler

return

fire basket

ashpan

glazed firedoor

tiled hearth

(c) Typical solid fuel boiler

Figure 3.45   **Boilers**
continued

(1)   An adequate flue shall be provided wherever necessary.
(2)   All materials and components shall comply with the requirements of the appropriate standard.
(3)   The use of an inadequate or badly maintained flue can have fatal consequences.

Circulating pumps

Pump circulation is needed in all cases where natural circulating pressure is insufficient (see figure 3.46).

Immersed rotor (glandless) type circulating pumps must be used on primary circuits only. Pumps used for secondary circulation must be resistant to corrosion.

Inlet and outlet connections should be fitted with full-way valves, and space allowed for renewal or repair.

Circulating pumps should be quiet in operation and suitably suppressed to prevent radio or television interference.

An electrical isolating switch must be fixed adjacent to and within sight of the circulating pump. The whole of the wiring, earthing, etc., must be

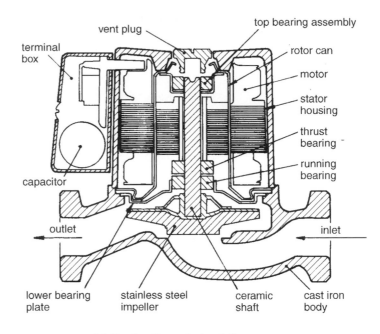

(a) Section through circulating pump

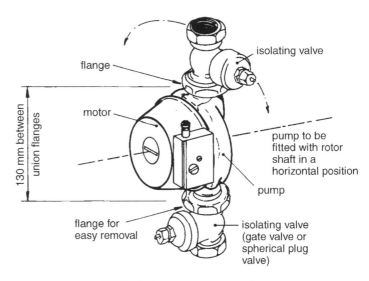

(b) Typical pump arrangement

Figure 3.46   Circulating pump

carried out by a competent electrician in accordance with BS 7671 (Requirements for Electrical Installations – IEE Wiring Regulations).

Circulating pumps should comply with the requirements of BS EN 60335-2-51, and should be installed in accordance with manufacturers' recommendations.

Valves and taps

Draw off taps and combination taps should comply with the requirements of BS 1010: Part 2 or BS 5412.

Valves used for isolating a section of the water service should provide a positive seal when closed. See also servicing valves, in section 2.4.

Pressure operated, temperature operated and combined relief valves, check valves, pressure reducing valves, anti-vacuum valves and pipe interrupters should be fitted in accordance with Building Regulations and Water Regulations (see chapters 4 and 5).

Draining taps should comply with the requirements of BS 1010 or BS 2879 (see figure 7.8), and be suitable for hosepipe connection.

Water Regulations require draining taps to be fitted so that the entire water system can be drained down when not in use, thus avoiding frost damage.

A further advantage is that systems can be readily drained before repair work is carried out so reducing the risk of mess and damage from pipes and fittings that are disconnected (see also chapter 7).

Mixing valves

Valves used for the mixing of hot and cold water (see figures 3.47–3.49) must be connected so that protection is provided against the possible contamination of water supplies. Where a mixer is used to control the water to more than one outlet, e.g. showers, it should be of the thermostatic type. Reference should also be made to the Health and Safety Executive Note (G) 104 'Safe' hot water and surface temperatures.

Tap mixers or combination taps

There are two types of tap mixers:

(1) Those that mix water within the valve body. These should be supplied from the same source, i.e. both hot and cold from a common storage cistern, or both under direct mains pressure. This ensures that no

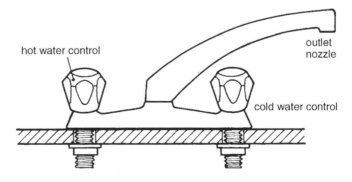

Figure 3.47   **Sink mixer tap**

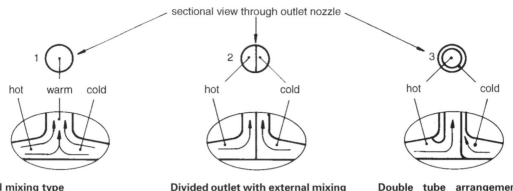

Internal mixing type

Mixing occurs in body of valve.

Precautions required against crossflow.

Difficult to maintain temperature where hot and cold inlet pressures vary.

Divided outlet with external mixing

Prevents crossflow between hot and cold supplies.

Some risk of scalding from hot outflow.

Double tube arrangement with external mixing

Prevents crossflow provided cold is on the outside and hot on the inside.

Reduced risk of scalding from outflow.

Figure 3.48   Types of mixer tap

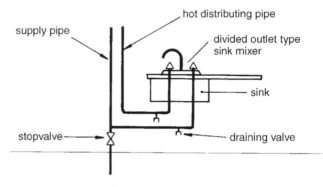

Hot and cold supplies from mixed sources may be connected as shown if the mixer is of the divided outlet type.

Where water is blended within the mixer, supplies should be from a common source. However, the above arrangement may be permitted if the hot water is supplied from a cistern complying with Water Regulations.

Figure 3.49   Sink mixer installation

crossflow can occur between water sources, and gives better and safer temperature and flow control.

(2)   Twin outlet types that mix water outside the valve outlet. These are of two designs. First, the divided outlet which creates no crossflow risk and can be connected to differing sources, and second, the type with a tube within a tube which is suitable for supply from separate sources, provided care is taken to ensure that the cold is on the outside, and the hot on the inside (see figure 3.48). This also reduces the risk of scalding.

Shower mixers

There are two types: thermostatic and non-thermostatic. Thermostatic types (see figure 3.50) are preferred because they provide automatic temperature control. They sense changes in temperature and adjust hot and cold flows accordingly. They also provide the user with a degree of protection against the effects of irregular flow and pressure variations caused when other taps are used. Pipework arrangements to showers are shown in figure 3.51.

High usage shower mixers supplying multiple shower outlets may need to be notified to the water undertaker before installation.

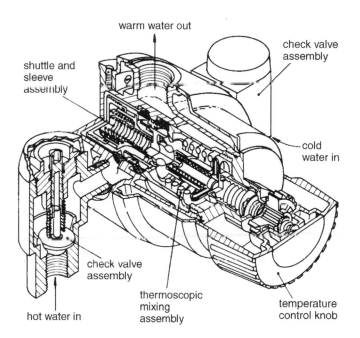

Thermoscopic (thermostatic) assembly provides temperature balance.

Shuttle and sleeve assembly will shut off hot water in the event of cold water failure.

Figure 3.50   Thermostatic shower mixer

3.8   Energy supply installations

The wiring to electric heaters must be in accordance with BS 7671 (requirements for Electrical Installations – IEE Wiring Regulations), and installed by a competent person such as a certificate holder of the National Inspection Council for Electrical Installation Contracting (NICEIC).

All gas installation work must comply with the Gas Safety (Installation and Use) Regulations and be carried out in accordance with relevant Codes of Practice. Installers must be registered with the Council for the Registration of Gas Installers (CORGI). Any person working on gas must be

(a) Storage-fed shower

Separate supplies to shower:
- ○ decrease risk of flow and pressure variations;
- ○ eliminate backflow risk.

Shower mixer supplied from one source, e.g. hot and cold from storage or both hot and cold direct from mains.

With a fixed shower head which has an air gap maintained at all times, there is no backflow risk.

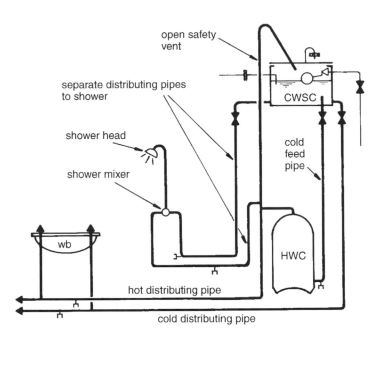

(b) Mains-fed shower

Where shower hose is constrained and type AA air gap is maintained at nozzle at all times, backflow prevention devices such as the double check valve assembly are not needed.

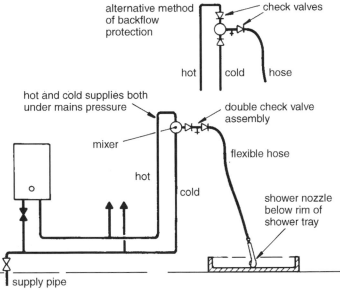

Figure 3.51   Shower installation

competent in gas work and carry a registration card showing he is competent in the area of gas work he is undertaking.

Solid fuel installations should be installed in accordance with the relevant Codes of Practice by a person such as a solid Fuel Advisory Service registered heating contractor.

Chapter 4
Prevention of bursting

It is extremely dangerous to heat water in a filled enclosed vessel to a temperature above 100°C. Water boils at 100°C at normal atmospheric pressure, but at 3 bar pressure, for example, the boiling point rises to 143°C and the heated water will have expanded and/or its pressure will have increased. If the water expands to such an extent that a small split develops in the cylinder, there will be an immediate drop in pressure and the water will flash to steam increasing in volume by about 1600 times. This will result in the total rupture or explosion of the vessel.

Building Regulations require that the temperature of stored hot water shall not exceed 100°C.

Successful and safe operation of a hot water system depends on the following:

- the right equipment, properly installed and maintained, and not exposed to misguided interference;
- reliability and durability of safety devices and equipment;
- location and choice of system and components;
- Kitemarked equipment, or equipment approved by the Water Research Centre, or where appropriate, unvented systems and safety devices approved under recognized schemes of approval and certification, e.g. EOTA (European Organisation for Technical Approvals) or UKAS (United Kingdom Accreditation Service);
- good maintenance, for continued safety, with reasonable expectation that this will continue;
- good control of temperature and pressure;
- efficient control and use of energy.

4.1 Energy control

Three independent forms of energy control are accepted by the Building Regulations approved documents: thermostatic control, temperature relief, and heat dissipation.

Thermostatic control

Effective thermostatic control is needed to prevent the temperature of stored water from rising above the normal expected hot water temperature of 60°C to 65°C. This may be achieved by using either a cylinder thermostat (see figure 4.1) or an immersion heater thermostat (see figure 4.2) depending on the methods used to heat the water. Thermostatic control is

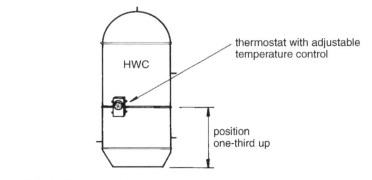

Figure 4.1   Cylinder thermostat

The permanent magnet ensures positive making and breaking of the contacts. This avoids arcing with consequent burned contacts and possible fire risk.

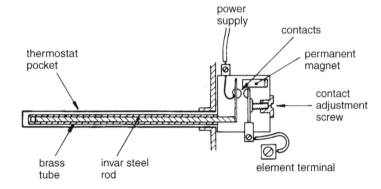

Figure 4.2   Section through immersion heater thermostat

used for the normal control of heat in hot water storage vessels, boilers and heating appliances (see figures 4.3 and 4.4.)

Thermostatic control should normally have the effect of shutting off the energy source (heat) immediately the design temperature is reached, and will, of course, allow the heat to be switched back on when the stored water temperature is lowered through use. There are, however, two exceptions:

(1)   In fully pumped primary circuits where a low water content boiler is used and the manufacturer recommends the use of a pump over-run to dissipate any excess heat generated by the boiler after the boiler is switched off.

(2)   In solid fuel boiler installations where the temperature of the stored water is controlled directly by the temperature of the water in the boiler. As the residual heat in the boiler is slow to die down, the system should have adequate capacity to absorb any excess heat generated during the cooling period. For this reason systems heated by solid fuel boilers are required to have a minimum capacity of 100 l hot water storage.

Heat cannot be shut off immediately as fuel takes time to cool down.

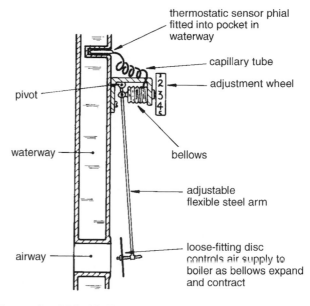

Figure 4.3 **Thermostatic control for small solid fuel boiler**

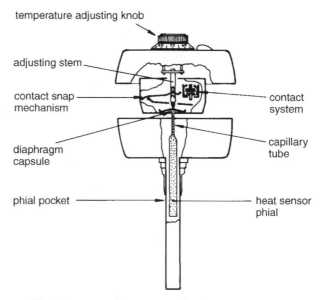

Used in gas or oil burners to give instant control.

Figure 4.4 **Boiler thermostat, liquid filled**

Temperature-operated energy cut-out device

This device, which may also be known as a 'thermal cut-out' device, is used in unvented systems and will only operate if the normal thermostat fails and the hot store vessel overheats. It must be of the non-self-resetting type, be independent of the thermostat and be designed to cut off the heat source at a predetermined temperature of 90°C. There must be a separate energy cut-out device for each heat source.

In directly heated systems, with an immersion heater to provide the heat source, the energy cut-out will be situated within the thermostat. Heaters having two or more immersion heaters should have independent cut-out devices within each immersion heater. Immersion heaters used as supplementary heaters to indirectly heated systems should have their own cut-out device independent of any fitted to control the heat from the boiler.

The indirectly heated system will usually have its energy cut-out device arranged to shut off a motorized valve on the flow to the cylinder, or directly to shut off the boiler, or alternatively there may be an energy cut-out device located on the boiler itself. Whichever method is employed, it is important that the water in the cylinder is not allowed to reach boiling point, which is the purpose of the energy cut-out device.

Temperature-operated energy cut-out devices are illustrated in figure 4.5.

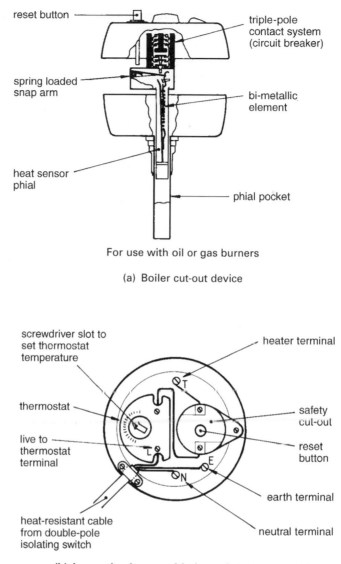

reset button

triple-pole contact system (circuit breaker)

spring loaded snap arm

bi-metallic element

heat sensor phial

phial pocket

For use with oil or gas burners

(a) Boiler cut-out device

screwdriver slot to set thermostat temperature

heater terminal

thermostat

safety cut-out

live to thermostat terminal

reset button

earth terminal

heat-resistant cable from double-pole isolating switch

neutral terminal

(b) Immersion heater with thermal cut-out devices

Figure 4.5 **High energy cut-out devices**

Temperature relief and heat dissipation

Adequate means of dissipating the heat input are needed in case both the temperature thermostat and the energy cut-out fail. Two methods are accepted by Building Regulations:

(1) A **vent pipe** to atmosphere, often known as the open safety vent, seen as part of the traditional vented hot water system, and shown in figures 3.25 and 3.26.

(2)   A **temperature relief valve** (see figure 4.6) as used for many years in unvented systems throughout the world. This must be fitted in the top of the storage vessel, within the top 20% volume of the water. This device is used as a safety back-up in case both the temperature thermostat and the thermal cut-out device fail.

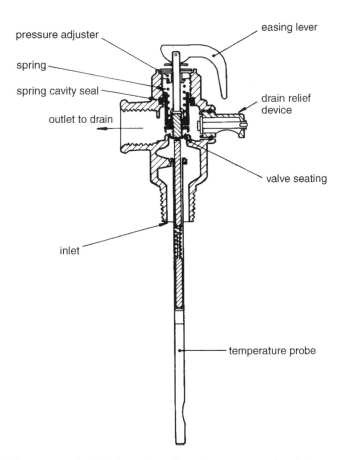

Discharge capacity 1.5 times that of maximum energy input to heater.

No valve to be fitted between temperature relief valve and heater.

Figure 4.6   **Temperature relief valve to BS 6283: Part 2**

The water discharged from a temperature relief valve must be removed from the point of discharge to a safe place. In operation the valve will open at a pre-set temperature to permit the overheated water to escape safely from the hot water storage vessel before it boils. It will usually operate at about 95°C.

Maintenance and periodic easing of temperature relief valves is particularly important for continued efficiency. A notice drawing attention to this should be provided in a prominent position for the user.

Sealed primary circuits

When sealed primaries are used; it is permissible for a second temperature-operated energy cut-out to be employed in place of the temperature relief valve. Sealed primary circuits must not be heated by solid fuel. Both a vent and a temperature relief valve must be fitted in addition to a thermostat if water in a primary circuit or in a direct system is heated by solid fuel. This is because complete thermostatic control and an effective temperature-operated energy cut-out are not available.

Discharges from temperature relief valves and expansion/pressure relief valves

Building Regulations, in Approved Document G3, state 'there shall be precautions to ensure that the hot water discharged from safety devices is safely conveyed to where it is visible but will cause no danger to persons in or about the building'.

Two important points are stated here.

(1) Any discharge from a temperature relief valve or expansion relief valve must be readily visible. This will:
    (a) show there is a fault on the system that requires maintenance;
    (b) reduce wastage of water because faults will be seen and rectified.
(2) Any discharge must be to a safe place. This will apply more importantly to the temperature relief valve, which will discharge very hot water from the top of the hot storage vessel, whereas the discharge from the pressure relief valve situated on the cold supply pipe will be relatively cold. See figures 4.7 and 4.8.

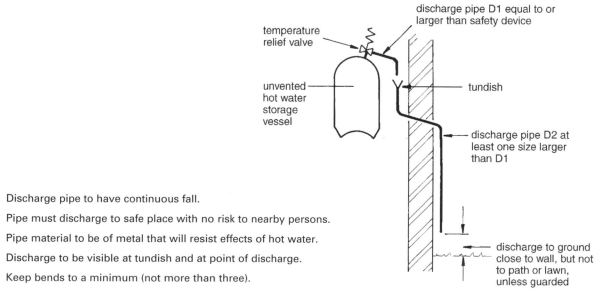

Discharge pipe to have continuous fall.

Pipe must discharge to safe place with no risk to nearby persons.

Pipe material to be of metal that will resist effects of hot water.

Discharge to be visible at tundish and at point of discharge.

Keep bends to a minimum (not more than three).

Figure 4.7   **Discharge from temperature relief valve**

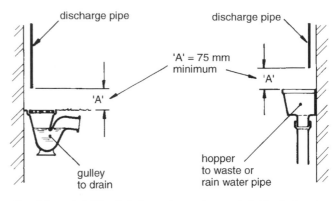

Consider suitability of drain or down pipe material for hot water.

Guidance document G3 to the Building Regulations shows the discharge pipe terminating below the gulley grating but above the trap water level. In this case the discharge will not be readily visible.

(a) To gulley **(b) To RWP hopper**

Figure 4.8 **Alternative methods of discharge**

The use of a tundish (see figure 4.9) will permit greater flexibility in the positioning of appliances and discharge outlet and will also provide visibility where discharges are more than 9 m from the appliance.

Discharges from expansion or pressure relief valves can be treated a little more leniently than those from temperature relief valves as the danger from the discharge is considerably less.

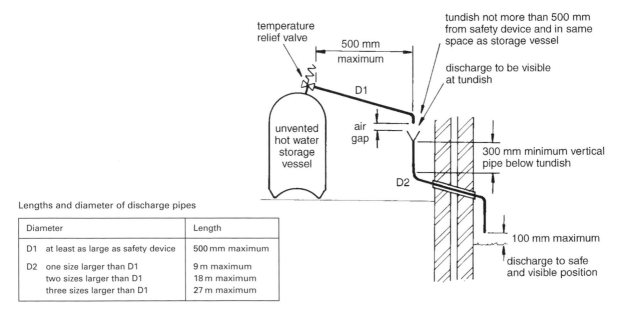

Lengths and diameter of discharge pipes

Diameter	Length
D1 at least as large as safety device	500 mm maximum
D2 one size larger than D1	9 m maximum
two sizes larger than D1	18 m maximum
three sizes larger than D1	27 m maximum

Figure 4.9 **Positioning and design of tundish**

Additional considerations for multiple installations, e.g. flats

The problems here are ensuring that visibility is maintained and also knowing which appliance is at fault should a discharge occur, (see figure 4.10).

Not more than six appliances to discharge to any one common discharge pipe

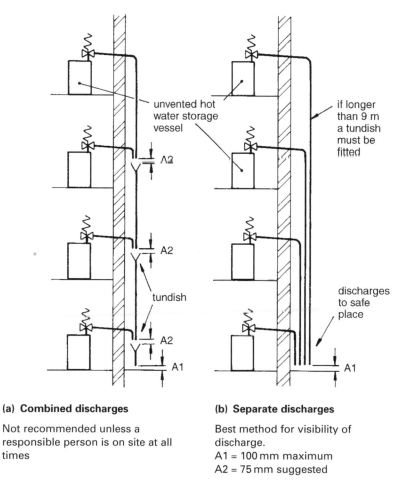

unvented hot water storage vessel

A2

A2

tundish

A2

A1

if longer than 9 m a tundish must be fitted

discharges to safe place

A1

(a) Combined discharges

Not recommended unless a responsible person is on site at all times

(b) Separate discharges

Best method for visibility of discharge.
A1 = 100 mm maximum
A2 = 75 mm suggested

Figure 4.10   Discharges from multiple installations

Pipework may be within the building, provided that tundishes and discharge terminals are readily visible and adequate provision is made to remove any discharges from the building safely, e.g. by trapped connections to drains.

Discharges from a temperature relief valve and an expansion relief valve may be combined in a common discharge pipe, see figure 4.11.

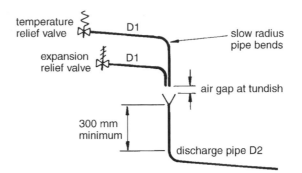

Discharge to safe and visible position.

Figure 4.11   **Combined discharge from temperature relief valve and expansion relief valve**

Sequence of operation

Thermostats, temperature-operated cut-outs and temperature relief valves must be set to operate in this sequence as the temperature increases. These three devices are not essential where water is only heated indirectly by a primary circuit which is already protected, or from a source of heat that is incapable of raising the temperature above 90°C.

4.2   Pressure and expansion control

Working pressures

In any hot water system the working pressure must not exceed the safe working pressure of the component parts, (see table 4.1).

Table 4.1   **Safe working pressures**

Circuit	Maximum working pressure bar	Test pressure bar
Sealed primary	3	5
Unvented secondary	6	10

Where necessary the supply pressure should be controlled by using a break cistern or pressure reducing valve, (see figures 4.12 and 4.13).

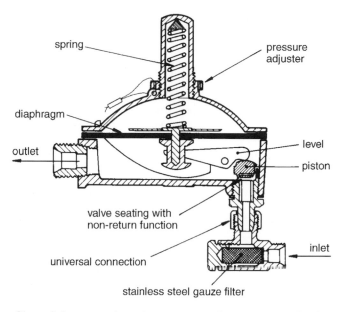

Gives finite control, and constant outlet pressure allowing accurate sizing of system.

Essential where copper hot water cylinder is used.

Figure 4.12   Diaphragm-actuated pressure reducing valve, to BS 6283: Part 4

Provides cruder control than the diaphragm-actuated reducing valve, to maintain supply line at preset maximum pressure.

Use only with glass-lined steel (or similar) cylinder on higher pressure system (up to 6 bar).

Not for use with copper cylinder on low pressure system (up to 3 bar).

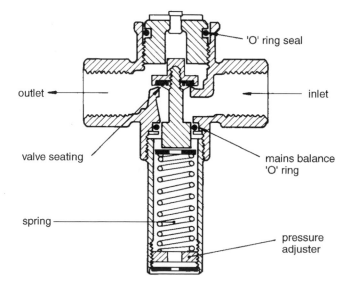

Figure 4.13   Pressure limiting valve

Control of water expansion

Within any hot water system the expansion of the heated water and consequent possible pressure rise must be limited without any discharge of water to waste. That is, expansion water must be accommodated within the hot water system. There are a number of ways of doing this depending upon the system (vented or unvented) and the size of the water heater and/ or any storage vessel.

(1) In the traditional vented system expanded water is permitted to move back into the cold feed pipe towards the feed cistern (see figure 4.14) but only as long as the cold feed pipe is unobstructed so that there will be no resistance to the flow of expanded water. Consequently, there will be no increase in pressure in the supply due to expansion.

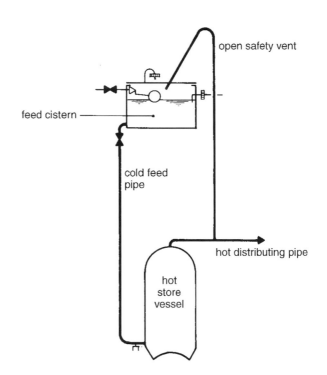

Feed cistern also serves as break cistern to limit pressure in system.

Valve not to have loose jumper.

Cold feed not to be fitted with check valve or other obstruction.

Open safety vent to permit escape of steam in the event of system overheating.

Vent to be open at all times.

Heated water expands safely into cold feed pipe.

Figure 4.14   Expansion in a traditional vented system

(2) In the unvented hot water storage system expansion control is normally achieved by using an expansion vessel or a dedicated air space within the hot store vessel, which should be rated to accommodate a volume of expanded water at least equal to 4% of the total volume of water likely to be heated in the system. (See figures 4.15 and 4.16.) The expansion vessel should be connected on the cold water inlet to the storage vessel. A check valve should be fitted to prevent backflow of expansion water into the cold feed pipe.

(3) Reversed flow into supply pipe. In any mains supplied unvented hot water supply system whether instantaneous, water-jacketed tube or storage type, reverse flow along the supply pipe may be permitted (see figure 4.17), provided that there is no restriction on the supply pipe such as a check valve, pressure reducing valve or stopvalve which might prevent the reversal of flow as and when water expands. The supply pipe should be large enough to accommodate a volume of water at least equal to 4% of the total volume of water to be heated. In these circumstances, no heated or warm water should reach either the communication pipe or any branch pipe feeding a cold water outlet.

This method, although permitted by Water Regulations, is not widely practised, particularly in systems installed in older properties where the water undertaker's stopvalve might incorporate a loose jumper.

(4) An expansion relief valve (see figure 4.18) may be used as a fail-safe device. However, it is not permitted to be used as the sole means of control of expansion water. This valve should be arranged to open automatically and to discharge water only when the pressure in the system reaches a predetermined level under failure conditions.

Any water discharged from an expansion valve must be discharged safely to a conspicuous position in a similar fashion to that of the temperature relief valve (see section 4.1).

Expansion vessel to conform to BS 6144.

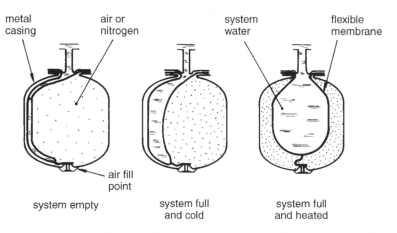

metal casing   air or nitrogen   system water   flexible membrane

air fill point

system empty   system full and cold   system full and heated

Flexible membrane prevents contact between water and steel casing to minimize corrosion.

Figure 4.15   Expansion vessel

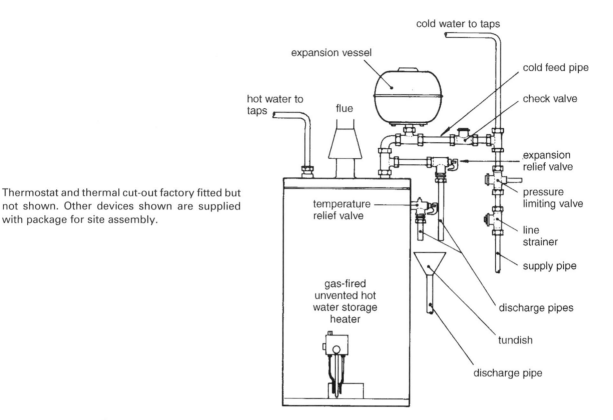

Thermostat and thermal cut-out factory fitted but not shown. Other devices shown are supplied with package for site assembly.

Figure 4.16   Use of expansion vessel in packaged unvented hot water system

Not recommended because of the possibility that an obstruction to supply pipe could be fitted at a later date.

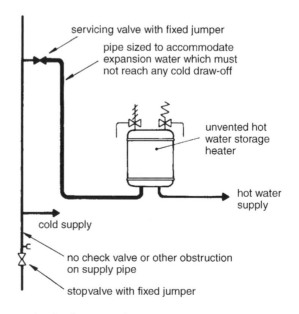

Figure 4.17   System using supply pipe for expansion

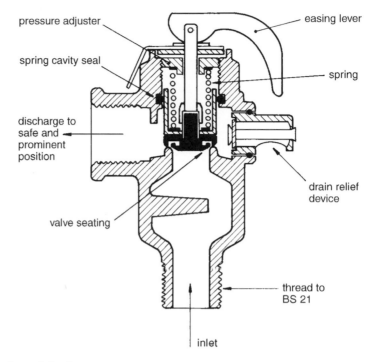

Figure 4.18   Expansion relief valve

4.3   Control of water level

Connections should be made so that water cannot be drawn (1) from any primary or closed circuit, and (2) to ensure that the hottest water can be drawn only from the top of a hot water storage vessel and above any primary flow connection or heating element. (See figure 4.19.) This includes the position of connections from secondary returns.

Unintentional draining down of any system (particularly storage types) is dangerous and should be avoided, as it may:

- expose temperature sensing controls, thus impairing their operation,
- expose the heating element, which could then become overheated,
- result in steam being produced with possibly disastrous consequences.

Vented primary circuits should be fitted with an adequate supply of make-up water, i.e. cold feed.

Sealed primary circuits, having no permanent supply of make-up water, should have a notice displayed drawing attention to the need for regular inspection and maintenance to keep the system and its pressure at the design level. Sealed systems must be under complete thermostatic control.

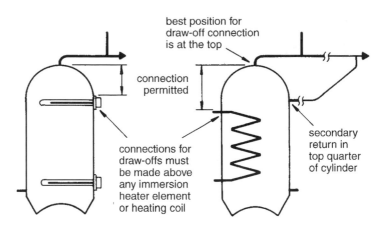

best position for
draw-off connection
is at the top

connection
permitted

secondary
return in
top quarter
of cylinder

connections for
draw-offs must
be made above
any immersion
heater element
or heating coil

(a) Draw-off connections to hot water cylinder

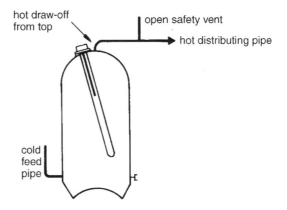

hot draw-off
from top

open safety vent

hot distributing pipe

cold
feed
pipe

(b) Cylinder with top entry heater element

Where immersion heater or primary coil is inserted at top of
cylinder any draw-off connection must be above immersion
heater connection. The only exception is the use of a draining
valve which must have a removable key for operation.

Figure 4.19   Control of water level

Chapter 5
Pipe sizing

Pipes and fittings should be sized so that the flow rates for individual draw-offs are equal to the design flow rates shown in table 5.1. During simultaneous discharges, flows from taps should not be less than the minimum flow rates shown in figure 5.1.

Flow velocities generally should not exceed 3 m/s. Filling times for cisterns may range from 1 to 4 hours depending on their capacity and the flow rate available from the local water supply. In single dwellings the filling time should not exceed 1 hour.

Correct pipe sizes will ensure adequate flow rates at appliances and avoid problems caused by oversizing and undersizing, see figure 5.1.

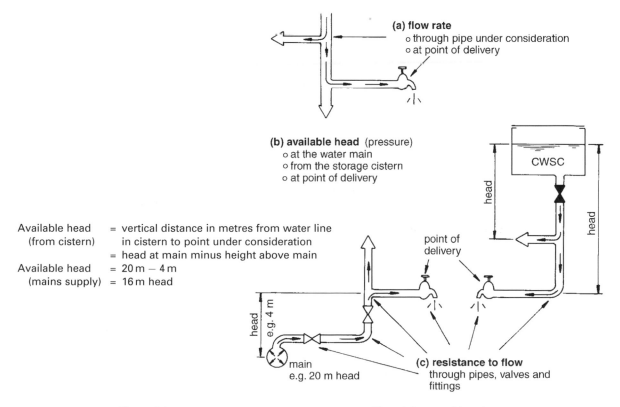

(a) flow rate
o through pipe under consideration
o at point of delivery

(b) available head (pressure)
o at the water main
o from the storage cistern
o at point of delivery

CWSC

head

head

Available head (from cistern) = vertical distance in metres from water line in cistern to point under consideration
= head at main minus height above main
Available head (mains supply) = 20 m − 4 m
= 16 m head

point of delivery

head e.g. 4 m

main e.g. 20 m head

(c) resistance to flow through pipes, valves and fittings

Figure 5.1   **Pipe sizing considerations**

Oversizing will mean:

- additional and unnecessary installation costs;
- delays in obtaining hot water at outlets;
- increased heat losses from hot water distributing pipes.

Undersizing may lead to:

- inadequate delivery from outlets and possibly no delivery at some outlets during simultaneous use;
- some variation in temperature and pressure at outlets, especially showers and other mixers;
- some increase in noise levels.

In smaller, straightforward installations such as single dwellings, pipes are often sized on the basis of experience and convention.

In larger and more complex buildings, or with supply pipes that are very long, it is necessary to use a recognized method of calculation such as that shown in sections 5.1 and 5.2.

5.1   Sizing procedure for supply pipes

The procedure below is followed by an explanation of each step with appropriate examples.

(1)   Assume a pipe diameter.
(2)   Determine the flow rate:
    (a)   by using loading units;
    (b)   for continuous flows;
    (c)   obtain the design flow rate by adding (a) and (b).
(3)   Determine the effective pipe length:
    (d)   work out the measured pipe length;
    (e)   work out the equivalent pipe length for fittings;
    (f)   work out the equivalent pipe length for draw-offs;
    (g)   obtain the effective pipe length by adding (d), (e) and (f).
(4)   Calculate the permissible loss of head:
    (h)   determine the available head:
    (i)   determine the head loss per metre run through pipes;
    (j)   determine the head loss through fittings;
    (k)   calculate the permissible head loss.
(5)   Determine the pipe diameter:
    (l)   Decide whether the assumed pipe size will give the design flow rate in (c) without exceeding the permissible head loss in (k).

Explanation of the procedure

Assume a pipe diameter

In pipe sizing it is usual to make an assumption of the expected pipe size and then prove whether or not the assumed size will carry the required flow.

Determine the flow rate

In most buildings it is unlikely that all the appliances installed will be used simultaneously. As the number of outlets increases the likelihood of them all being used at the same time decreases. Therefore it is economic sense to design the system for likely peak flows based on probability theory using loading units, rather than using the possible maximum flow rate.

(a) *Loading units.* A loading unit is a factor or number given to an appliance which relates the flow rate at its terminal fitting to the length of time in use and the frequency of use for a particular type and use of building (probable usage). Loading units for various appliances are given in table 5.1.

By multiplying the number of each type of appliance by its loading unit and adding the results, a figure for the total loading units can be obtained. This is converted to a design flow rate using figure 5.2.

An example using loading units is given in figure 5.3.

(b) *Continuous flows.* For some appliances, such as automatic flushing cisterns, the flow rate must be considered as a continuous flow instead of applying probability theory and using loading units. For such appliances the full design flow rate for the outlet fitting must be used, as given in table 5.1.

However, in the example shown in figure 5.3, the continuous flow for the two urinals of 0.008 l/s (from table 5.1) is negligible and can be ignored for design purposes.

(c) *Design flow rate.* The design flow rate for a pipe is the sum of the flow rate determined from loading units (a) and the continuous flows (b).

Table 5.1 **Design flow rates and loading units**

Outlet fitting	Design flow rate l/s	Minimum flow rate l/s	Loading units
WC flushing cistern single or dual flush	0.13	0.01	2
WC trough cistern	0.15 per WC	0.01	2
Wash basin tap size $\frac{1}{2}$ – DN 15	0.15 per tap	0.01	$1\frac{1}{2}$ to 3
Spray tap or spray mixer	0.05 per tap	0.03	—
Bidet	0.20 per tap	0.01	1
Bath tap, nominal size $\frac{3}{4}$ – DN 20	0.30	0.20	10
Bath tap, nominal size 1 – DN 25	0.60	0.40	22
Shower head (will vary with type of head)	0.20 hot or cold	0.10	3
Sink tap, nominal size $\frac{1}{2}$ – DN 15	0.20	0.10	3
Sink tap, nominal size $\frac{3}{4}$ – DN 20	0.30	0.20	5
Washing machine size $\frac{1}{2}$ DN 15 dishwasher size $\frac{1}{2}$ – DN 15	0.20 hot or cold 0.15	0.15 0.10	3
Urinal flushing cistern	0.004 per position served	0.002	—

Notes
(1) Flushing troughs are advisable where likely use of WCs is more than once per minute. For peak flows use 3 l/s.
(2) Mixer fittings use less water than separate taps, but this can be disregarded in sizing.
(3) Urinal demand is very low and can usually be ignored. Alternatively, use the continuous flow.
(4) Loading units should not be used for outlet fittings having high peak demands, e.g. those in industrial installations. In these cases use the continuous flow.

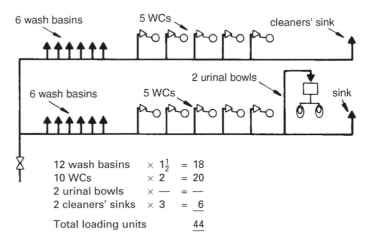

12 wash basins	× $1\frac{1}{2}$	=	18
10 WCs	× 2	=	20
2 urinal bowls	× —	=	—
2 cleaners' sinks	× 3	=	6
Total loading units			44

Therefore, from figure 5.2, the required flow rate for the system is 0.7 l/s.

Figure 5.2 **Conversion chart – loading units to flow rate**

Figure 5.3 **Example of use of loading units**

Determine the effective pipe length

(d) *Find the measured pipe length*. Figure 5.4 is an example showing how the measured pipe length is found.

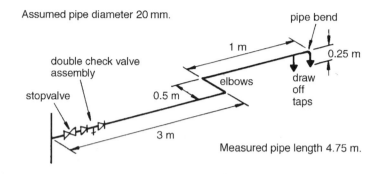

Note There is no need to consider both branch pipes to taps.

Figure 5.4   **Example of measured pipe length**

(e) and (f) *Find the equivalent pipe lengths for fittings and draw-offs*. For convenience the frictional resistances to flow through fittings are expressed in terms of pipe lengths having the same resistance to flow as the fitting. Hence the term 'equivalent pipe length' (see table 5.2).

For example, a 20 mm elbow offers the same resistance to flow as a 20 mm pipe 0.8 m long.

Figure 5.5 shows the equivalent pipe lengths for the fittings in the example in figure 5.4.

(g) *Effective pipe length*. The effective pipe length is the sum of the measured pipe length (d) and the equivalent pipe lengths for fittings (e) and draw-offs (f).

Therefore, for the example shown in figure 5.4 the effective pipe length would be:

Measured pipe length		4.75 m
Equivalent pipe lengths		
elbows	$2 \times 0.8 =$	1.6 m
tee	$1 \times 1.0 =$	1.0 m
stopvalve	$1 \times 7.0 =$	7.0 m
taps	$2 \times 3.7 =$	7.4 m
check valves	$2 \times 4.3 =$	8.6 m
Effective pipe length	$=$	30.35 m

Table 5.2   Equivalent pipe lengths (copper, stainless steel and plastics)

Bore of pipe mm	Equivalent pipe length			
	Elbow m	Tee m	Stopvalve m	Check valve m
12	0.5	0.6	4.0	2.5
20	0.8	1.0	7.0	4.3
25	1.0	1.5	10.0	5.6
32	1.4	2.0	13.0	6.0
40	1.7	2.5	16.0	7.9
50	2.3	3.5	22.0	11.5
65	3.0	4.5	—	—
73	3.4	5.8	34.0	—

Notes
(1) For tees consider change of direction only. For gate valves losses are insignificant.
(2) For fittings not shown, consult manufacturers if significant head losses are expected.
(3) For galvanized steel pipes in a small installation, pipe sizing calculations may be based on the data in this table for equivalent nominal sizes of smooth bore pipes. For larger installations, data relating specifically to galvanized steel should be used. BS 6700 refers to suitable data in the *Plumbing Engineering Services Design Guide* published by the Institute of Plumbing.

Using the example from figure 5.4:

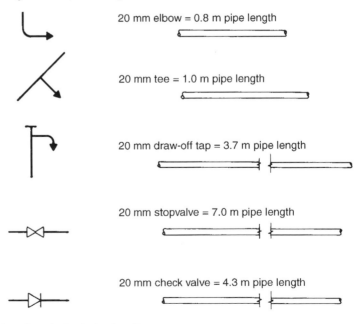

20 mm elbow = 0.8 m pipe length

20 mm tee = 1.0 m pipe length

20 mm draw-off tap = 3.7 m pipe length

20 mm stopvalve = 7.0 m pipe length

20 mm check valve = 4.3 m pipe length

Figure 5.5   Examples of equivalent pipe lengths

Permissible loss of head (pressure)

Pressure can be expressed in the following ways.

(i)   In pascals, the pascal (Pa) being the SI unit for pressure.
(ii)   As force per unit area, N/m^2.
    $1\,N/m^2 = 1$ pascal (Pa).
(iii)   As a multiple of atmospheric pressure (bar).
    Atmospheric pressure $= 100\,kN/m^2 = 100\,kPa = 1$ bar.
(iv)   As metres head, that is, the height of the water column from the water
    level to the draw-off point.
    $1\,m$ head $= 9.81\,kN/m^2 = 9.81\,kPa = 98.1\,mb$.

   In the sizing of pipes, any of these units can be used. BS 6700 favours
the pascal. However, this book retains the use of metres head, giving a
more visual indication of pressure that compares readily to the height and
position of fittings and storage vessels in the building.

(h)   *Available head.* This is the static head or pressure at the pipe or fitting
    under consideration, measured in metres head (see figure 5.1).
(i)   *Head loss through pipes.* The loss of head (pressure) through pipes
    due to frictional resistance to water flow is directly related to the length
    of the pipe run and the diameter of the pipe. Pipes of different
    materials will have different head losses, depending on the roughness
    of the bore of the pipe and on the water temperature. Copper,
    stainless steel and plastics pipes have smooth bores and only pipes of
    these materials are considered in this section.
(j)   *Head loss through fittings.* In some cases it is preferable to subtract
    the likely resistances in fittings (particularly draw-offs) from the avail-
    able head, rather than using equivalent pipe lengths.
       Table 5.3 gives typical head losses in taps for average flows com-
    pared with equivalent pipe lengths. Figures 5.6 and 5.7 provide a
    method for determining head losses through stopvalves and float-
    operated valves respectively.

    Note: Where meters are installed in a pipeline the loss of head
    through the meter should be deducted from the available head.

(k)   *Permissible head loss.* This relates the available head to the frictional
    resistances in the pipeline. The relationship is given by the formula:

$$\text{Permissible head loss (m/m run)} = \frac{\text{Available head (m)}}{\text{Effective pipe length (m)}}$$

This formula is used to determine whether the frictional resistance in a
pipe will permit the required flow rate without too much loss of head
or pressure. Figure 5.8 illustrates the permissible head loss for the
example in figure 5.4.

Table 5.3   **Typical head losses and equivalent pipe lengths for taps**

Nominal size of tap	Flow rate l/s	Head loss m	Equivalent pipe length m
G $\frac{1}{2}$ – DN 15	0.15	0.5	3.7
G $\frac{1}{2}$ – DN 15	0.20	0.8	3.7
G $\frac{3}{4}$ – DN 20	0.30	0.8	11.8
G 1 – DN 25	0.60	1.5	22.0

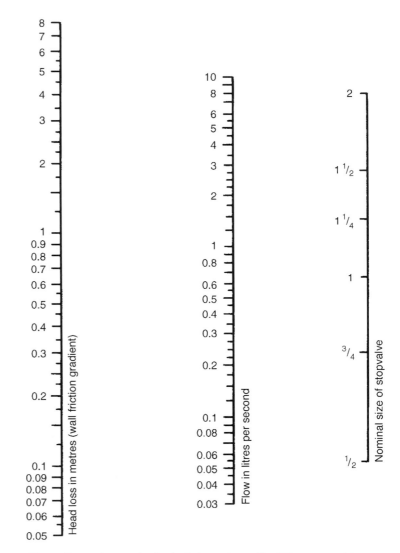

Note   Gate valves and spherical plug valves offer little or no resistance to flow provided they are fully open.

Figure 5.6   **Head loss through stopvalves**

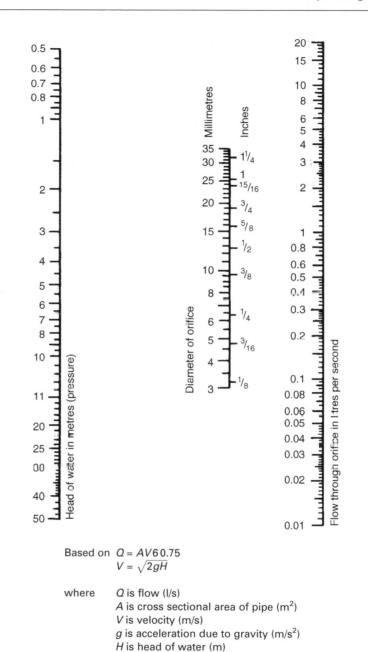

Based on $Q = AV6\,0.75$

$V = \sqrt{2gH}$

where Q is flow (l/s)
          A is cross sectional area of pipe (m^2)
          V is velocity (m/s)
          g is acceleration due to gravity (m/s^2)
          H is head of water (m)

Figure 5.7   **Head loss through float-operated valves**

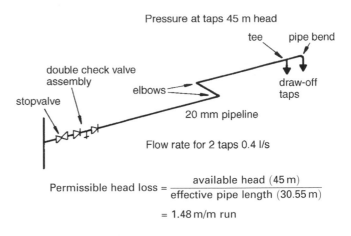

$$\text{Permissible head loss} = \frac{\text{available head (45 m)}}{\text{effective pipe length (30.55 m)}}$$

$$= 1.48\,\text{m/m run}$$

Figure 5.8   **Example of permissible head loss**

Determine the pipe diameter

In the example in figure 5.4 a pipe size of 20 mm has been assumed. This pipe size must give the design flow rate without the permissible head loss being exceeded. If it does not, a fresh pipe size must be assumed and the procedure worked through again.

Figure 5.9 relates pipe size to flow rate, flow velocity and head loss. Knowing the assumed pipe size and the calculated design flow rate, the flow velocity and the head loss can be found from the figure as follows.

(i)   Draw a line joining the assumed pipe size (20 mm) and the design flow rate (0.4 l/s).
(ii)   Continue this line across the velocity and head loss scales.
(iii)   Check that the loss of head (0.12 m/m run) does not exceed the calculated permissible head loss of 1.48 m/m run.
(iv)   Check that the flow velocity (1.4 m/s) is not too high by referring to table 5.4.

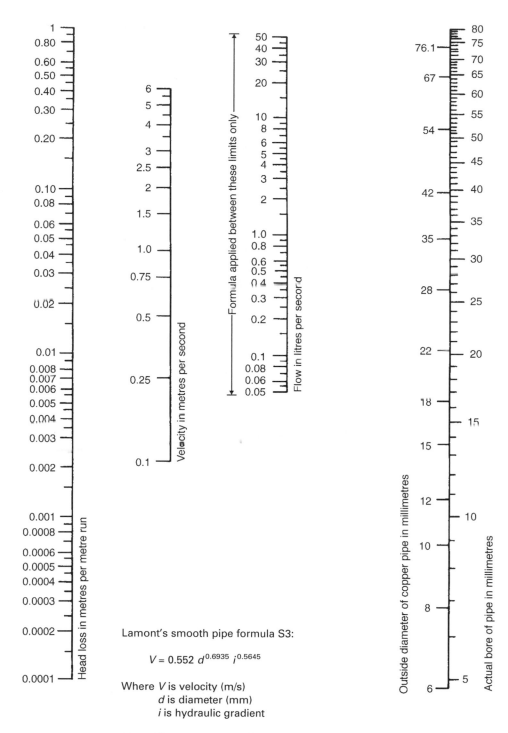

Lamont's smooth pipe formula S3:

$$V = 0.552\, d^{0.6935}\, i^{0.5645}$$

Where V is velocity (m/s)
d is diameter (mm)
i is hydraulic gradient

Note   Figures shown are for cold water at 12°C. Hot water will show slightly more favourable head loss results.

Figure 5.9   Determination of pipe diameter

Table 5.4   **Maximum recommended flow velocities**

Water temperature °C	Flow velocity	
	Pipes readily accessible m/s	Pipes not readily accessible m/s
10	3.0	2.0
50	3.0	1.5
70	2.5	1.3
90	2.0	1.0

Note   Flow velocities should be limited to reduce system noise.

5.2   Tabular method of pipe sizing

Pipe sizing in larger and more complicated buildings is perhaps best done by using a simplified tabular procedure. BS 6700 gives examples of this but for more detailed data readers should refer to the Institute of Plumbing's *Plumbing Engineering Services Design Guide*.

The data used in the tabular method that follows are taken from BS 6700 but the author has simplified the method compared with that given in the standard.

The tabular method uses a work sheet which can be completed as each of the steps are followed in the pipe sizing procedure. An example of the method follows with some explanation of each step.

Explanation of the tabular method

Pipework diagram

(1)   Make a diagram of the pipeline or system to be considered (see figure 5.10).
(2)   Number the pipes beginning at the point of least head, numbering the main pipe run first, then the branch pipes.
(3)   Make a table to show the loading units and flow rates for each stage of the main run. Calculate and enter loading units and flow rates, see figure 5.10.

Calculate flow demand

(1)   Calculate maximum demand (see figure 5.10):
 • add up loading units for each stage (each floor level);
 • convert loading units to flow rates;
 • add up flow rates for each stage.

(2)   Calculate probable demand (see figure 5.10):
 • add up loading units for all stages;
 • convert total loading units to flow rate.

(3)   Calculate percentage demand (number of stages for which frictional resistances need be allowed). See figure 5.12.

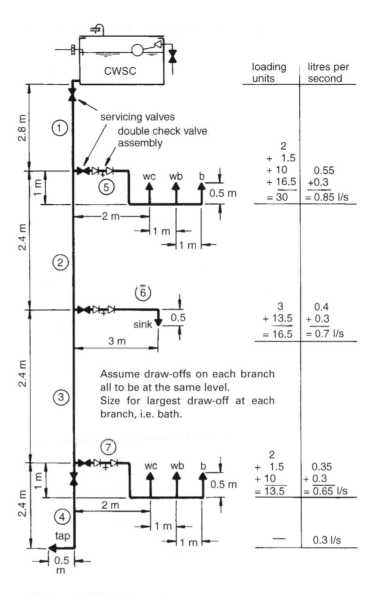

Bib tap at 0.3 l/s in frequent use.

Note Figure is not to scale for convenience, water level in cistern taken to be at base of cistern. Servicing valves assumed to be full-flow gate valves having no head losses.

Refer also to figure 5.12.

Figure 5.10   Pipe sizing diagram

Work through the calculation sheet

See figure 5.11, using the data shown in figures 5.10 and 5.12.

This is an example of a suitable calculation sheet with explanatory notes.

Calculation sheet

(1) Pipe reference	(2) Loading units	(3) Flow rate (l/s)	(4) Pipe size (mm diameter)	(5) Loss of head (m/m run)	(6) Flow velocity (m/s)	(7) Measured pipe run (m)	(8) Equivalent pipe length (m)	(9) Effective pipe length (m)	(10) Head consumed (m)	(11) Progressive head (m)	(12) Available head (m)	(13) Final pipe size (mm)	(14) Remarks
Enter pipe reference on calculation sheet	Determine loading unit (table 5.1)	Convert loading units to flow rate (figure 5.2)	Make assumption as to pipe size (inside diameter)	Work out frictional resistance per metre (figure 5.9)	Determine velocity of flow (figure 5.9)	Measure length of pipe under consideration	Consider frictional resistances in fittings (table 5.2 and figures 5.6 and 5.7)	Add totals in columns 7 and 8	Head consumed – multiply column 5 by column 9	Add head consumed in column 10 to progressive head in previous row of column 11	Record available head at point of delivery	Compare progressive head with available head to confirm pipe diameter or not	Notes

Note If, for any pipe or series of pipes, it is found that the assumed pipe size gives a progressive head that is in excess of the available head, or is noticeably low, it will be necessary to repeat the sizing operation using a revised assumed pipe diameter.

Figure 5.11   Calculation sheet – explanation of use

Estimated maximum demand $= 1.4$ l/s

Probable demand $= 0.85$ l/s

$$\text{Percentage demand} = \frac{\text{Probable demand}}{\text{estimated maximum demand}} \times \frac{100}{1}$$

$$= \frac{0.85}{1.4} \times \frac{100}{1} = 60\%$$

Therefore only 60% of the installation need be considered.

For example, if we were designing for a multi-storey building 20 storeys high, only the first 12 storeys need to be calculated.

However, in the example followed here, the whole system has been sized because the last fitting on the run has a high flow rate in continuous use.

For branches only the pipes to the largest draw-off, i.e. the bath tap, need be sized.

Calculation sheet

(1) Pipe reference	(2) Loading units	(3) Flow rate (l/s)	(4) Pipe size (mm diameter)	(5) Loss of head (m/m run)	(6) Flow velocity (m/s)	(7) Measured pipe run (m)	(8) Equivalent pipe length (m)	(9) Effective pipe length (m)	(10) Head consumed (m)	(11) Progressive head (m)	(12) Available head (m)	(13) Final pipe size (mm)	(14) Remarks
1	30	0.85	32	0.05	1.2	2.8	1.4	4.2	0.21	0.21	2.8	32	
5	13.5	0.35	20	0.095	1.25	5.5	12.0	17.5	1.66	1.87	3.3	20	
2	16.5	0.7	25	0.12	1.5	2.4	–	2.4	0.29	2.16	5.2	25	
6	3	0.3	20	0.07	1.0	3.5	10.4	13.9	0.97	3.13	5.7	20	
3	13.5	0.65	25	0.1	1.4	2.4	–	2.4	0.24	3.37	7.6	25	
7	13.5	0.35	20	0.095	1.25	5.5	12.0	17.5	1.66	5.03	8.1	20	
4	–	0.3	20	0.07	1.0	2.9	1.6	4.5	0.31	5.34	10.0	20	

Refer also to figure 5.10.

Figure 5.12   Calculation sheet – example of use

5.3   Sizing cold water storage

In Britain, cold water has traditionally been stored in both domestic and non-domestic buildings, to provide a reserve of water in case of mains failure. However, in recent years we have seen an increase in the use of 'direct' pressure systems, particularly for hot water services where many combination boilers and unvented hot water storage vessels are now being installed.

BS 6700 is inconsistent about the quantity of storage required in houses. In clause 2.2.3.11 it recommends:

Smaller houses	cistern supplying cold water only	– 100 l to 150 l
	cistern supplying hot and cold outlets	– 200 l to 300 l
Larger houses	per person where cistern fills only	– 80 l
	at night per person	– 130 l

However, in clause 2.3.9.4 it recommends a minimum storage capacity of 230 l where the cistern supplies both cold water outlets and hot water apparatus, which was a requirement of byelaws in the past. The author still favours the old byelaw requirements, which are more specific and which still seem to be the normal capacity installed.

Cold water storage cistern	– 115 l minimum
Feed cistern	– No minimum but should be equal to hot store vessel supplied
Combined feed and storage cistern	– 230 l minimum

For larger buildings the capacity of the cold water storage cistern depends on:

- type and use of buildings;
- number of occupants;
- type and number of fittings;
- frequency and pattern of use;
- likelihood and frequency of breakdown of supply.

These factors have been taken into account in table 5.5, which sets out minimum storage capacities in various types of building to provide a 24-hour reserve capacity in case of mains failure.

Table 5.5   **Recommended minimum storage of hot and cold water for domestic purposes**

Type of building	Minimum cold water storage litres (l)	Minimum hot water storage litres (l)
Hostel	90 per bed space	32 per bed space
Hotel	200 per bed space	45 per bed space
Office premises:		
with canteen		
facilities	45 per employee	4.5 per employee
without canteen		
facilities	40 per employee	4.0 per employee
Restaurant	7 per meal	3.5 per meal
Day school:		
nursery } primary }	15 per pupil	4.5 per pupil
secondary } technical }	20 per pupil	5.0 per pupil
Boarding school	90 per pupil	23 per pupil
Children's home or		
residential nursery	135 per bed space	25 per bed space
Nurses' home	120 per bed space	45 per bed space
Nursing or convalescent home	135 per bed space	45 per bed space

Note   Minimum cold water storage shown includes that used to supply hot water outlets.

Example of calculation of minimum water storage capacity

Determine the amount of cold water storage required to cover 24 hours interruption of supply in a combined hotel and restaurant. Number of hotel guests 75, number of restaurant guests 350.

Storage capacity = number of guests × storage per person from table 5.5

Hotel storage capacity $= 75 \times 200 = 15\,000\,l$

Restaurant storage capacity $= 350 \times 7 = 2450\,l$

Therefore total storage capacity required is $15\,000\,l + 2450\,l = 17\,450\,l$

5.4   Sizing hot water storage

Minimum hot water storage capacities for dwellings, from BS 6700, are:

- 35 l to 45 l per occupant, unless the heat source provides a quick recovery rate,
- 100 l for systems heated by solid fuel boilers,
- 200 l for systems heated by off-peak electricity.

The feed cistern should have a capacity at least equal to that of the hot storage vessel.

The standard gives little information on storage capacities for larger buildings so in table 5.5 the author has given data for this based on that in the Institute of Plumbing's *Plumbing Engineering Services Design Guide*. The calculations are similar to those in section 5.3 for the minimum cold water storage capacity in larger buildings, and require the number of people to be multiplied by the storage per person.

BS 6700 takes a different approach to the sizing of hot water storage and suggests that when sizing hot water storage for any installation, account must be taken of the following:

- pattern of use;
- rate of heat input to the stored water (see table 5.6);
- recovery period for the hot store vessel;
- any stratification of the stored water.

Table 5.6   **Typical heat input values**

Appliance	Heat input kW
Electric immersion heater	3
Gas-fired circulator	3
Small boiler and direct cylinder	6
Medium boiler and indirect cylinder	10
Directly gas-fired storage hot water heater (domestic type)	10
Large domestic boiler and indirect cylinder	15

Stratification (see figure 3.42) means that the hot water in the storage vessel floats on a layer of cold feed water. This enables hot water to be drawn from the storage vessel without the incoming cold feed water mixing appreciably with the remaining hot water. In turn, this allows a later draw-off of water at a temperature close to the design storage temperature, with less frequent reheating of the contents of the storage vessel and savings in heating costs and energy.

Stratification is most effective when cylinders and tanks are installed vertically rather than horizontally, with a ratio of height to width or diameter of at least 2:1.

The cold feed inlet should be arranged to minimize agitation and hence mixing, by being of ample size and, if necessary, fitted with a baffle to spread the incoming water.

Stratification is used to good effect in off-peak electric water heaters (see figure 5.13). In this case no heat is normally added to the water during the daytime use and consequently very little mixing of hot and cold water takes place. In other arrangements the heating of the water will induce some mixing.

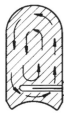

(a) Bottom entry heater

more even
temperature
throughout
cylinder

With a bottom entry immersion heater mixing will occur when
the water is being heated.

(b) Top entry heater

hot at top

warm

cold at bottom

With top entry immersion heater stratification will prevent
mixing

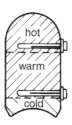

hot

warm

cold

(c) Twin entry immersion heater

With a twin entry immersion heater the top entry element can
provide economical energy consumption for normal use with
the bottom entry element operating when large quantities of
water are needed, for example, for bathing and washing.

With off-peak heaters the top element can be brought into use
during on-peak periods when needed to top up hot water, and
stratification will ensure that on-peak electricity is not used to
excess.

Figure 5.13   Effects of stratification

Calculation of hot water storage capacity

As noted above, the storage capacity required in any situation depends on
the rate of heat input to the stored hot water and on the pattern of use. For
calculating the required storage capacity BS 6700 provides a formula for
the time M (in min) taken to heat a quantity of water through a specified
temperature rise:

$$M = VT/(14.3P)$$

where

V is the volume of water heated (in l);
T is the temperature rise (in °C);
P is the rate of heat input to the water (in kW).

This formula can be applied to any pattern of use and whether stratification
of the stored water takes place or not. It ignores heat losses from the hot
water storage vessel, since over the relatively short times involved in

reheating water after a draw-off has taken place, their effect is usually small.

The application of this formula to the sizing of hot water cylinders is best illustrated by the following examples, in which figures have been rounded.

In these examples a small dwelling with one bath installed has been assumed. Maximum requirement: 1 bath (60 l at 60°C plus 40 l cold water) plus 10 l hot water at 60°C for kitchen use, followed by a second bath fill after 25 min.

Thus a draw-off of 70 l at 60°C is required, followed after 25 min by 100 l at 40°C, which may be achieved by mixing hot at 60°C with cold at 10°C.

Example 1   Assuming good stratification

Good stratification could be obtained, for example, by heating with a top entry immersion heater. With a rate of heat input of 3kW, the time to heat the 60 l for the second bath from 10°C to 60°C is:

$$M = VT/(14.3P)$$
$$M = (60 \times 50)/(14.3 \times 3)$$
$$M = 70 \, min$$

Since the second bath is required after 25 min, it has to be provided from storage. But in the 25 min the volume of water heated to 60°C is:

$$V = M \, (14.3)/T$$
$$V = (25 \times 14.3 \times 3)/50$$
$$V = 21 \, l$$

Therefore the minimum required storage capacity is:

$$70 + 6 - 21 = 109 \, l$$

Example 2   Assuming good mixing of the stored water

Good mixing of the stored water would occur, for example, with heating by a primary coil in an indirect cylinder.

Immediately after drawing off 70 l at 60°C for the first bath and kitchen use, the heat energy in the remaining water plus the heat energy in the 70 l replacement at 10°C equals the heat energy of the water in the full cylinder.

The heat energy of a quantity of water is the product of its volume and temperature. Then, if V is the minimum size of the storage cylinder and T is the water temperature in the cylinder after refilling with 70 l at 10°C:

$$(V - 70) \times 60 + (70 \times 10) = VT$$
$$T = (60V - 4200 + 700)/V$$
$$T = (60V - 3500)/V$$
$$T = 60 - 3500/V$$

The second bath is required after 25 min. Hence, with a rate of heat input of 3 kW:

$$25 = VT/(14.3 \times 3)$$

and the temperature rise $T = (25 \times 14.3 \times 3)/V$

and $T = 1072.5/V$

A temperature of at least 40°C is required to run the second bath. Therefore the water temperature of the refilled cylinder after the first draw-off of 70 l, plus the temperature rise after 25 min, must be at least 40°C, or:

$$(60 - 3500/V) + (1072.5V) = 40 \text{ (or more)}$$
$$60 - 2427.5/V = 40$$
$$20 = 2427.5/V$$
$$V = 122 \text{ l}$$

Table 5.7   **Hot water storage vessels – minimum capacities**

Heat input to water kW	Dwelling with 1 bath		Dwelling with 2 baths*	
	With stratification litres (l)	With mixing litres (l)	With stratification litres (l)	With mixing litres (l)
3	109	122	165	260
6	88	88	140	200
10	70	70	130	130
15	70	70	120	130

Note * Maximum requirement of 130 l drawn off at 60°C (2 baths plus 10 l for kitchen use) followed by a further bath (100 l at 40°C) after 30 min.

These calculations, which may be carried out for any situation, show the value of promoting stratification wherever possible. They also show the savings in storage capacity that can be made, without affecting the quality of service to the user, by increasing the rate of heat input to the water. Results of similar calculations are shown in table 5.7 and are taken from BS 6700.

5.5   *Legionella* – implications in sizing storage

It has been common practice in the past for water suppliers to recommend cold water storage capacities to provide for 24 hours of interruption in supply. Table 5.5 in this book and Table 1 of BS 6700 reflect this practice.

Recent investigations into the cause and prevention of *Legionella* contamination suggest that hot and cold water storage should be sized to cope with peak demand only, and that in the past, storage vessels have often been over-sized.

Reduced storage capacities would mean quicker turnover of water and less opportunity for *Legionella* and other organisms to flourish.

Stratification in hot water can lead to ideal conditions for bacteria to multiply, as the base of some cylinders often remains within the 20°C temperature range. The base of the cylinder may also contain sediment and hard water scale which provide an ideal breeding ground for bacteria.

Where, however, good mixing occurs, higher temperatures can be obtained at the cylinder base during heating up periods. This is helpful because *Legionella* will not thrive for more than 5 minutes at 60°C and are killed instantly at temperatures of 70°C or more.

Chapter 6
Preservation of water quality

Water undertakers in England and Wales have a duty under the Water Industry Act 1991 to provide a supply of wholesome water which is suitable and safe for drinking and culinary purposes. Similar conditions apply in Scotland and Northern Ireland. At the same time, public demand requires that water supplied is of good appearance with minimal colour, taste or odour.

To enable water undertakers to maintain their supplies in wholesome condition and to preserve the quality of water supplied, the Act provides water undertakers with powers to enforce Water Regulations which came into effect on 1 July 1999. Installers and users must ensure that systems and components which are installed and used comply with Water Regulations.

BS 6700 looks at the preservation of water quality in four main areas.

(1) *Materials in contact with water*. There is not much point in having a good quality water supply if it becomes contaminated by unsuitable pipes, fittings and jointing materials.
(2) *Stagnation of water* and the prevention of bacterial growth particularly at temperatures between 20°C and 50°C.
(3) *Cross connections* must be prevented between pipes supplied directly under mains pressure and pipes supplied from other sources, such as:
     (a) water from a private source
     (b) non-potable water
     (c) stored water
     (d) water drawn off for use.
(4) *Prevention of backflow* from fittings or appliances into services or mains, particularly at point of use, i.e. draw-off taps, flushing cisterns, washing machines, dishwashers, storage cisterns and hose connections. For example, pumps should not be connected so as to cause backflow into the supply pipe, (see figure 6.1).

Backflow prevention forms the largest portion of clause 2.6 of BS 6700 and is prominent in Paragraph 15 of Schedule 2 of the Water Regulations.

Where it is intended to use a pump to increase flow or pressure through a supply pipe, the water supplier should first be consulted and permission granted.

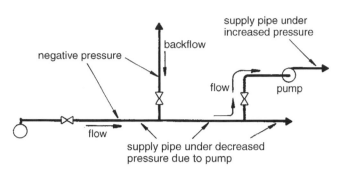

Pump may cause pressure drop sufficient to reverse flow in adjacent pipes.

Figure 6.1   **Example of backflow caused by pump installation**

6.1   Materials in contact with water

Contamination of water by contact with unsuitable materials will be avoided if careful attention during design and installation is given to:

- the specification and selection of acceptable materials used in the manufacture of pipes, fittings and appliances;
- the method of installation, and in particular to the method and materials used when jointing and connecting pipes, fittings and appliances;
- the environment into which pipes, fittings and appliances are to be installed;
- the design of the various elements of installation, especially where differing materials are to be used.

Materials selected for use in contact with water intended for domestic purposes should comply with Water Regulations and in particular Paragraph 2 of Schedule 2 'materials and substances in contact with water'. In general this means compliance is assured where any materials used are manufactured to a relevant BS or EN specification, or are listed in the *Water Fittings and Materials Directory* produced under the Water Regulations Advisory Scheme.

Examples of problems caused by unsuitable materials, and situations which should be avoided are shown in figures 6.2 and 6.4.

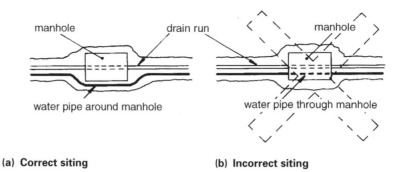

(a) **Correct siting**   (b) **Incorrect siting**

Pipes not to pass through any foul soil, refuse or refuse chute, ash pit, sewer or drain, cess pool or manhole.

Figure 6.2   Pipes not to pass through manholes, etc.

Pipes made of any material which is susceptible to permeation by gas or to deterioration by contact with substances likely to cause contamination of water should not be laid or installed in a place where permeation or deterioration is likely to occur. Because plastic pipes in particular are liable to a degree of permeation and deterioration by gas and oil, care should be taken when positioning pipelines to avoid contact in the event of leakages occurring from oil or gas lines, for example, at petrol filling stations. Figure 6.3 shows the positioning of pipelines in trenches as suggested by the *National Joint Utilities Group Report No. 6.*

Positioning should permit access to any one service without damage to another.

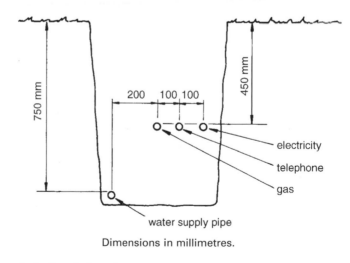

Dimensions in millimetres.

Figure 6.3   Positioning of pipelines in trenches

Substances leached from some materials may adversely affect the quality of water. Although British Standard schemes for the testing of pipes, fittings and materials aim to prevent this, much depends on the installer and the material chosen for a particular application, after taking due account of the nature of the water (see figure 6.4).

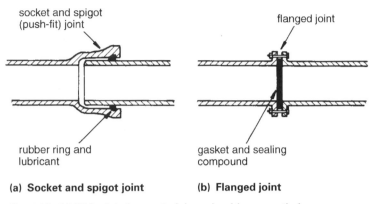

(a) Socket and spigot joint   **(b) Flanged joint**

See table 11.22 for jointing materials and guidance on their use.

The *Water Fittings and Materials Directory* produced under the Water Regulations Advisory Scheme lists materials and fittings which are approved for use.

Figure 6.4   **Examples of materials which could cause contamination if not chosen with care**

The use of lead for pipes, cisterns or in solders is prohibited because of the obvious danger of plumbo solvency. However, there will be cases when connections have to be made to existing lead pipes (see section 11.15).

For soldered joints to copper pipes, lead-free solders should be specified, e.g. tin/silver alloys.

The use of coal tar for the lining of pipes and cisterns is also prohibited. The direct connection between copper and lead pipes which might lead to electrolytic action and further lead solvency is prohibited in the absence of suitable means to prevent corrosion though galvanic action.

6.2 Stagnation of water and *Legionella*

All waters are likely to contain some bacterial infection. This should not be a problem in hot and cold water supplies as long as systems are properly installed and regularly well maintained. It is important that water in systems is kept clean and used quickly, with it not being allowed to lie in pipes, fittings and storage vessels for long periods.

To restrict bacterial growth in stored water, temperatures between 20°C and 50°C should be avoided. Water Regulations, in Paragraph 9 of Schedule 2, require that cold water used for domestic purposes is not warmed above 25°C.

This advice is particularly important in the control of Legionnaire's disease. Evidence for the occurrence of this disease has generally been found where pipes and components are not regularly maintained, or in parts of systems that cannot be cleaned easily during routine maintenance programmes, e.g. where flushing valves or draining valves are wrongly positioned, or where the accumulation of rust or other debris can settle and build up.

Legionnaire's disease is caused by *Legionella pneumophila*, a pneumonia-like microscopic bacterium that attacks the lungs. The bacteria occur naturally in water and can easily colonize hot and cold water supply systems unless preventative measures are taken. The disease can cause minor illness to healthy people but can prove to be fatal in those who are particularly vulnerable to respiratory diseases or those who are chronically ill. BS 6700 draws attention to the Health and Safety Commission's Approved Code of Practice which in turn draws attention to the following situations of particular concern:

(a)   hot water systems of more than 300 l capacity, and
(b)   hot and cold systems where occupants are particularly susceptible such as health care premises.

Because *Legionella* affects the lungs, and is caused by inhaling very small droplets of infected water, any location or fitting (e.g. showers, spray taps, humidifiers, etc.) that creates a water spray or aerosol mist should be given careful consideration before installation, and water supply systems should be designed, installed and maintained to avoid any risk.

Protective measures that should be taken to minimize colonization by *Legionella* and other harmful bacteria in the water are as follows:

- Cold water should be stored and distributed at as low a temperature as possible and preferably below 20°C.
- Hot water should be stored at 60°C to 65°C and distributed to provide delivery temperatures above 50°C after 1 minute.
- Materials that harbour or promote bacterial growth should be avoided.
- Fittings that tend to create aerosol formation should be avoided.
- Good throughput of water should be maintained.
- Storage capacities should be minimized to give adequate supplies for peak demand but to provide quick and regular replacement of stored water.
- New installations and newly repaired pipework and fittings should be flushed and sterilized before putting into use.
- Measures should be taken to prevent the accumulation of dirt, rust and other sediment within the water system, and where possible prevent them from gaining access to the system.
- Regular cleaning and maintenance of cisterns, pipes and fittings is essential.
- The inlet and outlet should be positioned at opposite ends of cold water storage cisterns.
- Drainage connections on larger cisterns should be positioned at the lowest part of the cistern and cisterns arranged with a fall to assist

draining down and cleaning.
- Storage vessels and pipelines that will have little use should be avoided. Otherwise an isolating valve should be fitted and appropriate backflow prevention devices considered.

For further information in a down-to-earth manner, the author recommends readers to the Institute of Plumbing's booklet *Legionnaire's Disease – Good Practice Guide for Plumbers*.

6.3   Prevention of contamination by cross connection

'Any water fitting conveying rainwater, recycled water or any fluid other than that supplied by the water undertaker, or any fluid that is not wholesome water, shall be clearly identified so as to be easily distinguished from any supply pipe or distributing pipe.'

'No supply pipe, distributing pipe or pump delivery pipe drawing water from a supply pipe shall convey or be connected so that it can convey any rainwater, recycled water or any fluid other than that supplied by the water undertaker, or any fluid that is not wholesome water.'

The two statements above set out the requirements of Paragraph 14 of Schedule 2 of the Water Regulations and quite simply mean that cross connection between wholesome water and unwholesome water is not permitted.

It is quite interesting that the regulations single out rainwater and recycled water. That these are given individual attention is a reflection of the government's view that both rainwater and recycled water could make a useful contribution to water conservation in the future. In its 'Recommendations for requirements to replace water byelaws' the Water Regulations Advisory Committee made the point that as the use of these waters is increased so is the risk of contamination by cross-connection and the risk of backflow is also likely to increase.

Apart from recycled water and rainwater, there are a number of 'unwholesome' waters that could give rise to contamination if cross-connections are made between them and pipes containing wholesome water. These include:

- water supplies and a reserve storage facility for fire fighting;
- water from private wells, springs, etc. (not supplied by the undertaker);
- stored water other than 'protected' drinking water;
- recycled water from industrial equipment or food production processes.

Examples of cross connections are illustrated in figure 6.5 whilst the correct installations for a variety of situations can be seen later in figures 6.27–6.29.

(a) Connections between supply pipe and distributing pipe

No connection shall be made between a supply pipe and a distributing pipe.

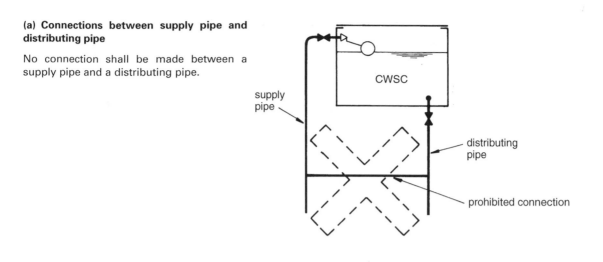

(b) Supplies from different sources

Installation not permitted. Correct installation is type AA air gap on supply pipe. See figure 6.7.

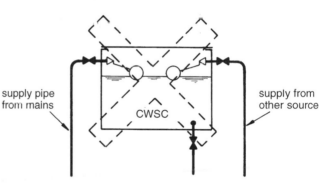

(c) Temporary connection to sealed heating system

No closed (sealed) circuit to be connected to a supply pipe unless adequate backflow prevention devices are in place.

Important because primary circuits to heating systems are likely to be contaminated with additives.

This diagram shows an approved method of protection for filling a sealed heating circuit in a house using a 'double check valve arrangement'.

Where the connection is to be a permanent arrangement, the Water Regulations Guide recommends the use of a 'type CA backflow preventor with different pressure zones'.

In premises other than a house when backflow risk is more severe, the minimum protection should be a 'type BA backflow preventor with reduced pressure zones' (RPZ valve).

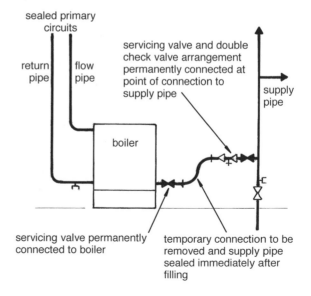

Figure 6.5   Cross connection hazards

(d) Shower mixer installation

No connection to be made between a supply pipe and a hot or cold distributing pipe.

Water could flow from hot store vessel to the drinking tap if the supply pipe should fail.

May be permitted if feed cistern is protected from contamination and a check valve is fitted to both the hot and cold supplies to the shower.

For correct installation see figure 3.51.

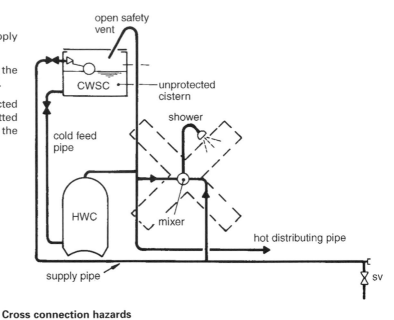

Figure 6.5   **Cross connection hazards**
continued

In order to avoid **any potential hazard** and ensure that hazardous connections are not made inadvertently, or otherwise, it is important that all pipes and fittings can be identified to show what is contained within them. Remember, Water Regulations require water fittings containing unwholesome water to be clearly identified. In practice it is best that *all* pipes can be distinguished one from another.

- All pipes should be colour coded to show their contents.
- Draw-off taps should be labelled to show those that are suitable for drinking purposes and those that are not.
- Valves and their function should be identified, particularly in industrial and commercial buildings where systems are more complex and unwholesome water more commonly used.
- Accurate drawings should be made, and passed over to the customer, showing where pipes are situated both within buildings and below ground so as to help distinguish them from one another.

See later illustrations, figures 11.66 and 11.67.

It is also important to remember that supply pipes and distributing pipes should not be connected together, even though they may both be supplying water for domestic purposes, unless in certain cases (e.g. shower mixers) appropriate backflow devices are fitted.

Additives to primary hot water or heating circuits

If a liquid (other than water) is used in any type of heating primary circuit, or if an additive, e.g. a corrosion inhibitor, is used in water in such a circuit, the liquid or additive should be non-toxic and non-corrosive.

6.4   Backflow protection

Backflow, which may include both 'backsiphonage' and 'back pressure', may be described as 'the reversal of the flow of water in a water supply system that can lead to contamination of the water supply'.

Backsiphonage may be described as 'backflow caused by siphonage of water from a cistern or appliance back into the pipe which feeds it'. For example, a hosepipe in use when the mains supply is turned off or a severe break in the main occurs causing water to run back into the supply pipe and/or the main.

Back pressure may be described as 'the reversal of flow in a pipe caused by an increase in pressure in the system'. An example of back pressure can be seen in the expansion of water from an unvented hot water heater which is permitted to pass back into the supply pipe.

Fluid risk categories

We have become familiar with three categories of backflow risk required by water byelaws which implemented the earlier recommendations of the Backsiphonage Report of 1974.

The Water Supply (Water Fittings) Regulations 1999 now recognize and list, in Schedule 1, five fluid categories based on those developed by the Union of Water Supply Associations of Europe (EUREAU 12) and which are also used in North America and Australia. The five fluid categories represent a range of water qualities depending on how 'drinkable' they are, or how much danger to health they might present.

A comparison between the current fluid categories and the former backflow risk categories is shown in figure 6.6 and a description of the five fluid categories is given in table 6.1 and is taken from Schedule 1 of the Regulations.

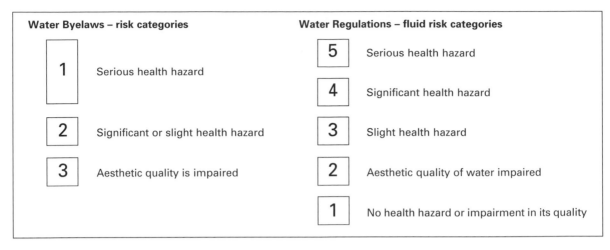

Figure 6.6 Comparison of backflow risk categories and fluid risk categories

Table 6.1 Fluid categories

Fluid category 1
Wholesome water supplied by a water undertaker and complying with the requirements of regulations made under Section 67 of the Water Industry Act 1991(a).

Fluid category 2
Water in fluid category 1 whose aesthetic quality is impaired owing to –
  (a) a change in its temperature, or
  (b) the presence of substances or organisms causing a change in taste, odour or appearance,
including water in a hot water distribution system.

Fluid category 3
Fluid which represents a slight health hazard because of the concentration of substances of low toxicity, including any fluid which contains –
  (a) ethylene glycol, copper sulphate solution or similar chemical additives, or
  (b) sodium hypochlorite (chloros and common disinfectants).

Fluid category 4
Fluid which represents a significant health hazard because of the concentration of toxic substances, including any fluid which contains –
  (a) chemical, carcinogenic substances or pesticides (including insecticides and herbicides), or
  (b) environmental organisms of potential health significance.

Fluid category 5
Fluid representing a serious health hazard because of the concentration of pathogenic organisms, radioactive or very toxic substances, including any fluid which contains –
  (a) faecal material or other human waste,
  (b) butchery or other animal waste, or
  (c) pathogens from any other source.

Backflow prevention

Appropriate preventative measures should:

- prevent any water returning from any appliance, fitting or process back into the supply or distributing pipe that feeds it, and
- prevent any water returning from a supply pipe back into the water undertaker's main.

It should be remembered that all water taken from a main, whether actually drawn off or still within the water system, is considered suspect and should not be permitted to return to the main.

However, water is permitted to flow back into a supply pipe or distributing pipe where the water has expanded from a hot water system or an instantaneous water heater, but the expanded water should be contained and not permitted to flow back to a position where it could be drawn off. Expansion in hot water vessels is discussed in chapter 4

Backflow can be avoided by good system design and the proper positioning of appropriate backflow prevention devices (see section 6.5). Backflow prevention devices should be placed as near to the point of use as is practically possible, and all devices should be accessible for examination, maintenance, renewal or repair. Non-mechanical means of protection are preferred wherever practicable and will include air gaps, tap gaps, and vented distributing pipes for cistern fed systems.

Guidance to Paragraph 2 of Schedule 2 of the Water Regulations gives the following advice on the installation of mechanical backflow devices:

- they should be readily accessible for inspection, maintenance and renewal;
- devices for protection against categories 2 and 3 should not be located outside buildings. Exceptions are noted for types HA and HUK1 devices that are made specifically for use with existing garden taps in domestic premises;
- devices, particularly vented or verifiable types or those with relief outlets, should not be installed below ground or in chambers below ground;
- where line strainers are installed, e.g. upstream of fluid category 4 prevention devices, a servicing valve should be fitted immediately before the line strainer;
- relief outlets for reduced pressure zone valves should terminate with a type AA air gap at least 300 mm above ground or floor level.

Backflow prevention measures, and devices used to prevent backflow should be related to the level of risk, and will vary according to the nature and use of the water supply.

6.5 Backflow prevention devices

A backflow prevention device is defined as 'a device which is intended to prevent contamination of drinking water by backflow'. It may be a mechanical or non-mechanical fitting or arrangement strategically positioned to prevent backflow from occurring.

There are a number of backflow prevention devices in various shapes and forms to suit to a range of fluid risk categories. The *Regulator's Specification on Prevention of Backflow* describes these in two tables, the first listing 10 non-mechanical arrangements, and the second showing 14 mechanical devices. The tables, which also indicate the fluid category for which each device is suited, are reproduced here in tables 6.2 and 6.3.

A selection of these arrangements and devices are described and illustrated on the following pages.

Table 6.2   **Schedule of non-mechanical backflow prevention arrangements and the maximum permissible fluid category for which they are acceptable**

Type	Description of backflow prevention arrangements and devices	Fluid category for which suited	
		Back pressure	Backsiphonage
AA	Air gap with unrestricted discharge above spill-over level	5	5
AB	Air gap with weir overflow	5	5
AC	Air gap with submerged inlet	3	3
AD	Air gap with injector	5	5
AF	Air gap with circular overflow	4	4
AG	Air gap with minimum size circular overflow determined by measure or vacuum test	3	3
AUK1	Air gap with interposed cistern (for example, a WC suite)	3	5
AUK2	Air gaps for taps and combination fittings (tap gaps) discharging over appliances, such as a wash basin, bidet, bath or shower tray shall not be less than the following: *size of tap or combination fitting* *vertical distance of bottom of tap outlet above spill-over level of receiving appliance* not exceed in $G\frac{1}{2}$ 20 mm exceeding $G\frac{1}{2}$ but not exceeding $G\frac{3}{4}$ 25 mm exceeding $G\frac{3}{4}$ 70 mm	X	3
AUK3	Air gaps for taps or combination fittings (tap gaps) discharging over any higher risk domestic sanitary appliance where a fluid category 4 or 5 is present, such as: (a) any domestic or non-domestic sink or other appliance; or (b) any appliance in premises where a higher level of protection is required, such as some appliances in hospitals or other health care premises, shall be not less than 20 mm or twice the diameter of the inlet pipe to the fitting, whichever is he greater	X	5
DC	Pipe interrupter with permanent atmospheric vent	X	5

Notes:

(1) X indicates that the backflow prevention arrangement or device is not applicable or not acceptable for protection against back pressure for any fluid category within water installations in the UK.

(2) Arrangements incorporating DC type devices shall have no control valves on the outlet side of the device; they shall be fitted not less than 300 mm above the spill-over level of a WC pan, or 150 mm above the sparge pipe outlet of a urinal, and discharge vertically downwards.

(3) Overflows and warning pipes shall discharge through, or terminate with, an air gap, the dimension of which should satisfy a type AA air gap.

Table 6.3 **Schedule of mechanical backflow prevention arrangements and the maximum permissible fluid category for which they are acceptable**

Type	Description of backflow prevention arrangements and devices	Fluid category for which suited	
		Back pressure	Backsiphonage
BA	Verifiable backflow preventer with reduced pressure zone (RPZ valve assembly)	4	4
CA	Non-verifiable disconnector with different pressure zones not greater than 10%	3	3
DA	Anti-vacuum valve (or vacuum breaker)	X	3
DB	Pipe interrupter with atmospheric vent and moving element	X	4
DUK1	Anti-vacuum valve combined with a single check valve	2	3
EA	Verifiable single check valve	2	2
EB	Non-verifiable single check valve	2	2
EC	Verifiable double check valve	3	3
ED	Non-verifiable double check valve	3	3
HA	Hose union backflow preventer. Only permitted for use on existing hose union taps in house installations	2	3
HC	Diverter with automatic return (normally integral with some domestic appliance applications only)	X	2
HUK1	Hose union tap which incorporates a double check valve. Only permitted for replacement of existing hose union taps in house installations	3	3
LA	Pressurized air inlet valve	X	2
LB	Pressurized air inlet valve combined with a check valve downstream	2	2

Notes:
(1) X indicates that the backflow prevention device is not acceptable for protection against back pressure for any fluid category within water installations in the UK.
(2) Arrangements incorporating a type DB type device shall have no control valves on the outlet side of the device. The device shall be fitted not less than 300 mm above the spill-over level of an appliance and discharge vertically downwards.
(3) Types DA and DUK1 shall have no control valves on the outlet side of the device and be fitted on a 300 mm minimum type A upstand.
(4) Relief outlet ports from types BA and CA backflow prevention devices shall terminate with an air gap, the dimension of which should satisfy a type AA air gap.

Non-mechanical backflow arrangements

Type AA air gap with unrestricted discharge

The type AA air gap satisfies the requirements of BS 6281: Part 1. It has become familiar under water byelaws but now has a new designation under Water Regulations. The type AA air gap gives high-risk protection (fluid category 5) from both backsiphonage and back pressure.

See figures 6.7 and 6.8. This is a pipe arrangement where the supply to a cistern or vessel is arranged as follows:

- the cistern or vessel has an unrestricted overflow to atmosphere;
- the supply discharge and its outlet are not obstructed;
- the water is discharged vertically or not more than 15° from the vertical;
- the vertical distance between the discharge pipe outlet and the spill-over level of the cistern or vessel is *not less* than that shown in table 6.4.

Its principal merit is that it has no moving parts and cannot readily be destroyed by vandals. It is accepted throughout the world.

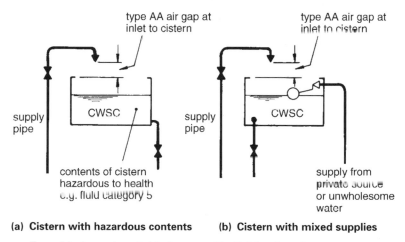

(a) Cistern with hazardous contents   **(b) Cistern with mixed supplies**

Type AA air gap is suitable for use with all risk categories.

Air gap related to size of inlet (see table 6.4).

Flow from inlet to be into air at atmospheric pressure and not more than 15° from the vertical.

Figure 6.7   **Examples of type AA air gap at cisterns**

Table 6.4   **Air gaps at taps, valves and fittings (including cisterns)**

Situation	Nominal size of inlet tap, valve or fitting	Vertical distance between tap or valve outlet and spill-over level of receiving appliance (mm)
Domestic situations fluid category 2/3 (device AUK2)	Up to and including $G\frac{1}{2}$ Over $G\frac{1}{2}$ and up to $G\frac{3}{4}$ Over $G\frac{3}{4}$	20 25 70
Non-domestic situations fluid category 4/5 (device AUK3)	Any size inlet pipe	Minimum diameter 20 mm, or twice the diameter of the inlet pipe, whichever is the greater

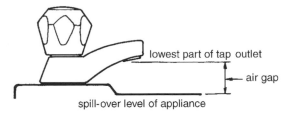

Type AA air gap to be visible, measurable and unobstructed.

Air gap is related to the size of the tap and the category of risk of the appliance (see table 6.4).

Figure 6.8   Example of type AA air gap at draw-off tap

Type AB air gap with weir overflow

The type AB air gap (see figure 6.9) is suitable for high-risk (category 5) situations and is particularly useful where there is a need to protect the contents of the storage vessel from ingress of contaminants such as dust and insects, e.g. feed and expansion cisterns in industrial or commercial premises.

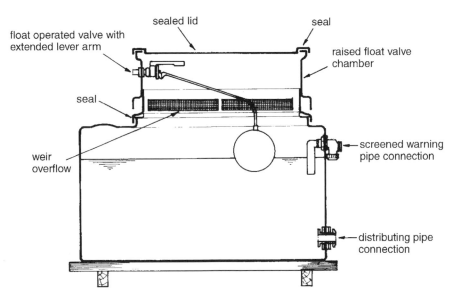

Weir overflow and warning pipe screened to prevent ingress by insects, dust, etc.

All joints on cistern sealed to prevent contamination.

Screened cistern vent, and open vent connection not shown.
These should be sealed to prevent contamination to the cistern water.

Figure 6.9   Example of type AB air gap with weir overflow

Type AG air gap with minimum size circular overflow

The type AG air gap arrangement satisfies the requirements of BS 6281: Part 2. See figure 6.10. In a cistern or other similar vessel, which is open at all times to the atmosphere, the vertical distance between the lowest point of discharge and the critical water level should be one of the following:

(1)   sufficient to prevent backsiphonage of water into the supply pipe;
(2)   not less than the distances shown in table 6.4.

In storage cisterns the type AG type air gap, as shown in figure 6.10, is impractical. Its critical water level, which determines the difference in fixing heights between the float-operated valve and the warning pipe, cannot be readily calculated. The critical water level will vary from one installation to another depending on the inlet pressure and the length and gradient of the warning pipe.

In most cases it is expected that water undertakers will accept the simpler arrangement seen in figure 6.11.

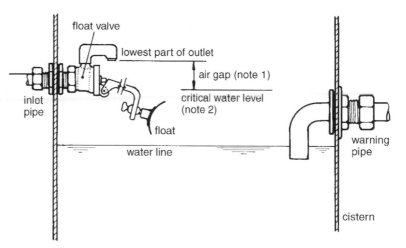

Note 1   The air gap is related to the size of the inlet and is the minimum permitted vertical distance between the 'critical' water level and the lowest part of the float valve outlet (see table 6.4).

Note 2   The critical water level is the highest level the water will reach at the maximum rate of inflow, i.e. float removed.

Note 3   Type AG air gaps should comply with the requirements of BS 6281: Part 2.

Figure 6.10   Type AG air gap (type B air gap to BS 6281: Part 2)

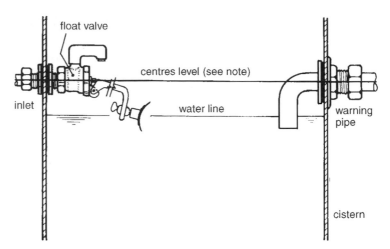

This arrangement is acceptable if:
- ○ the cistern complies with Water Regulations;
- ○ the float-operated valve is of the reducing flow type.

A reducing flow type float valve is one which gradually closes as the water level in the cistern rises, e.g. diaphragm float valve to BS 1212: Parts 2 or 3.

In this cistern the critical water level is assumed to be level with the centre line of the float valve body.

Figure 6.11   Acceptable alternative to the type AG air gap

The AUK1 air gap with interposed cistern

The AUK1 air gap with interposed cistern is used to 'disconnect' supply pipes and distributing pipes from appliances and vessels that might cause a backflow risk and is effective against:

(a)   backsiphonage risks of any fluid category; and
(b)   back pressure risk up to fluid category 3.

Common examples can be seen in WCs and urinals where the flushing cistern acts and in the interposed cistern (see figure 6.12). Further examples in commercial and industrial situations are shown later in this chapter.

Type AUK1 air gap with interposed cistern illustrated here in WC.

The AUK1 air gap should conform with type AG and comply with BS 6281: Part 2.

Float-operated valve should comply with BS 1212: Parts 2, 3 or 4

Dimension A 15 mm minimum

Dimension B 300 mm minimum

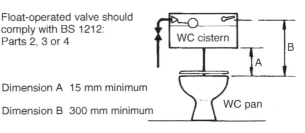

Figure 6.12   Type AUK1 air gap with interposed cistern

Type DC pipe interrupter with permanent atmospheric vent

A pipe interrupter (see figure 6.13) will admit air into a system, without the use of moving of flexible parts, to prevent backflow of water when a vacuum occurs. Whilst this device contains no moving parts, it can be subject to vandalism (by blockage of the airways). This device is suitable for protection from fluid category 5 backsiphonage risks, but cannot be used to protect against back pressure.

Typical use is with pressure flushing valves to WCs and urinals, where air is pulled in to break the siphon if a backflow condition occurs.

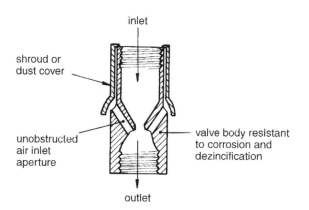

Valve should produce vacuum on outlet side and must discharge vertically downwards.

Must be fitted at least 300 mm above overflowing level of appliance served or 150 mm in the case of a urinal.

No tap or valve to be installed downstream (outlet size).

Pipe downstream not to be reduced in size.

Length of pipe downstream to be as short as possible.

Pipe interrupter to be readily accessible for repair.

Pipe interrupter to comply with BS 6281: Part 3.

Figure 6.13   **Example of pipe interrupter – type DC**

Mechanical backflow arrangements

Type BA verifiable backflow preventer with reduced pressure zone valve

The **reduced pressure zone valve (RPZ)**, as it is better known, is relatively new to the backflow scene and is designed to meet all but the very highest backflow risks, being rated to fluid category 4 for both back pressure and backsiphonage.

The type BA, RPZ valve is shown in figure 6.14 whilst installation details are shown in figure 6.15. Figure 6.16 shows how the valve works.

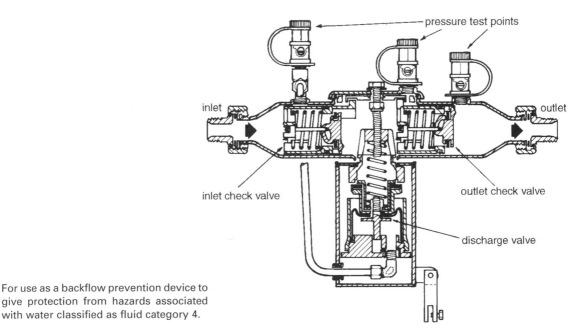

For use as a backflow prevention device to give protection from hazards associated with water classified as fluid category 4.

Figure 6.14   **Type BA, RPZ valve – sectional view**

Type CA non-verifiable disconnector with difference between pressure zones not greater than 10%

The non-verifiable disconnector works by venting the intermediate pressure zone of the valve to the atmosphere when the incoming pressure drops to 10% of the pressure at the outlet of the device. This valve will give protection against back pressure and backsiphonage risks up to fluid category 3.

Type DA anti-vacuum valve, or vacuum breaker

An anti-vacuum valve or vacuum breaker (see figure 6.17) is a mechanical device with an air inlet which remains closed when water flows past it, but which opens to admit air if there is a vacuum in the pipe. The vacuum breaker must close once pressure and flow return to normal.

Upstands are required for anti-vacuum valves to work to maximum effect, and provide additional protection should the protection at point of use fail (see figure 6.18).

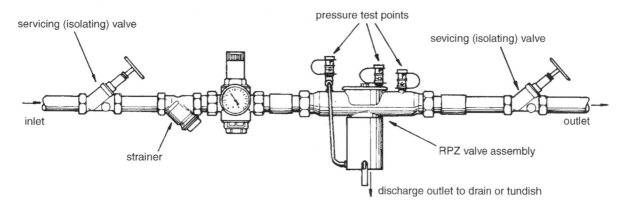

Install in acceptable position, free from flooding, free from effects of frost and in a secure, lockable cabinet.

Assembly to be installed above ground in horizontal position (unless otherwise approved).

Clearance to be allowed for test equipment to be fitted and to permit maintenance; minimum 100 mm at rear and 200 mm in front.

Adequate drainage to be provided, and air gap arranged between relief valve outlet and tundish/drain.

No drinking water supply to be drawn off after (downstream) of an RPZ assembly.

RPZ valve assemblies will cause a drop in pressure and may not be suitable for use on low pressure supplies.

Assemblies to be flushed out and sterilized after installation and before use.

Assembly to be checked following installation to ensure relief valves function correctly.

Assembly to be 'site tested' before use and at regular intervals after installation by an accredited tester and test certificate to be issued after each test.

Records should be kept of all installation and maintenance procedures. Copies to be retained by both installer and the customer.

Installers and testers may need to be registered with the water supplier.

Water supplier to be notified and approval given before RPZ valve is installed, and supplier should be notified of any maintenance tests carried out.

For more comprehensive information refer to *Information and guidance note No.9-03-02* issued by the WRC Evaluation and Testing Centre, or from the water supplier.

Figure 6.15   Example of RPZ valve installation

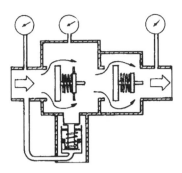 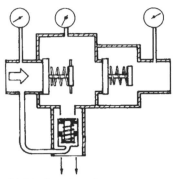 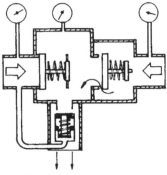

(a) Normal flow-through function

For normal function with flow-through valve, both check valves are open, and discharge value is closed.

(b) No flow function

Both check valves are closed and discharge valve is open.

(c) Backflow function

Upstream check valve is closed and downstream check valve is open. Discharge valve is open.

Figure 6.16   How the RPZ valve works

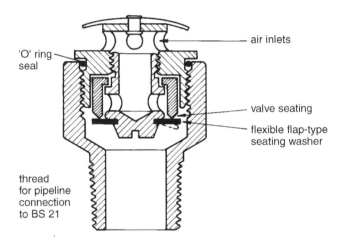

Air inlet remains closed when water inside the device is at or above atmospheric pressure, and opens to admit air under vacuum conditions at the valve inlet.

Diagram shows terminal type; in-line type also available.

Needs regular opening and shutting to keep it operable.

Must be same size as pipe to which it is connected.

No control valve to be fitted downstream.

Failure of these valves to close would mean waste of water and possible damage to building.

Satisfies Water Regulations if it complies with BS 6282: Part 2 and there is no control valve downstream.

Must be fitted at least 300 mm above overflow level of appliance served.

Figure 6.17   Type DA anti-vacuum valve, atmospheric type

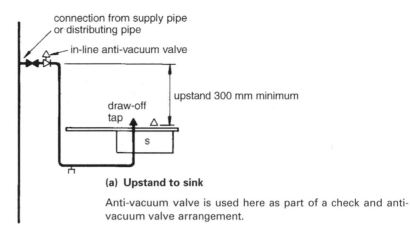

connection from supply pipe
or distributing pipe

in-line anti-vacuum valve

upstand 300 mm minimum

draw-off
tap

S

(a) Upstand to sink

Anti-vacuum valve is used here as part of a check and anti-vacuum valve arrangement.

Upstand is necessary for anti-vacuum valve to be effective.

Upstand 300 mm for cisterns and fixed appliances.

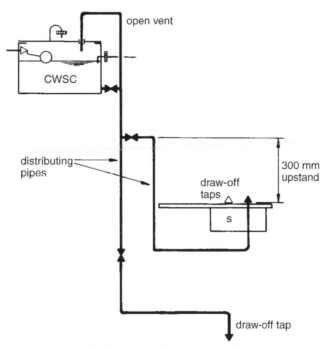

open vent

CWSC

distributing
pipes

300 mm
upstand

draw-off
taps

S

draw-off tap

(b) Upstand to sink, cistern fed

Open vent shown here is an alternative to the check and anti-vacuum valve.

Upstand is needed for the vent to be fully effective.

Where vent is used, upstand should be at least 300 mm.

Branch not to rise above its connection.

Figure 6.18   Upstands to anti-vacuum valves

continued

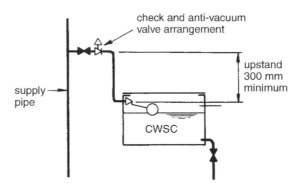

(c) Upstand to cistern

Upstand 300 mm above cistern and fixed appliances.

**Figure 6.18
continued**   **Upstands to anti-vacuum valves**

Type EB non-verifiable single check valve

A check valve (see figure 6.19) is a mechanical device which, by means of a resilient 'elastic' seal or seals, permits flow of water in one direction only and is closed when there is no flow. The check valve is suitable for low-risk fluid category 2 only. Typical use is on an inlet to a domestic water softener.

The check valve should:
○ be resistant to corrosion and dezincification;
○ operate satisfactorily at temperatures up to 65°C;
○ when shut, prevent any flow from inlet to outlet where water pressure does not exceed 10 mb;
○ should comply with BS 6282: Part 1.

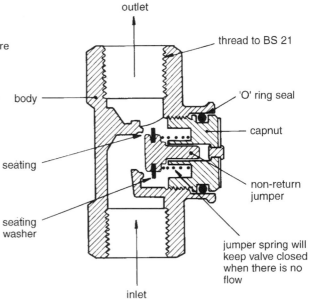

Figure 6.19   Check valve to BS 6282: Part 1

Type ED non-verifiable double check valve assembly

This comprises two check valves with a test cock fitted between them (see figure 6.20).

Double check valve assembly should comply with BS 6282: Part 1.

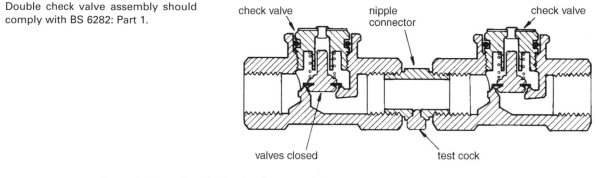

Figure 6.20   **Double check valve assembly**

Type DUK1 combined check and anti-vacuum valve

This is shown in figures 6.21 and 6.22 and consists of a single check valve upstream combined with an anti-vacuum valve. It is suitable for category 3 backsiphonage risk and category 2 back pressure risk.

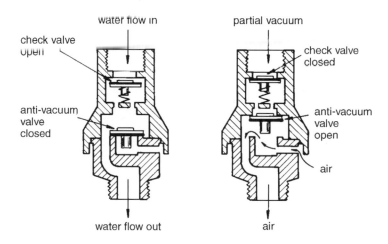

Must be used with an upstand for anti-vacuum valve to be effective.

Should comply with BS 6282: Part 4.

Figure 6.21   **Combined check and anti-vacuum valve assembly**

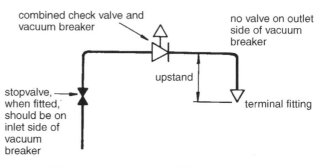

Upstand 300 mm above cistern and fixed appliances.

Check valve to be fitted upstream of vacuum breaker.

Figure 6.22   Use of combined check valve and anti-vacuum valve assembly

Categories of backflow risk

Table 6.5 lists appliances which fall into each of five fluid risk categories. The table lists minimum backflow protection measures, but a higher grade device may be used if more convenient. For situations not listed, the water undertaker should be consulted.

Table 6.5   Determination of fluid categories with examples

(a) Determination of fluid category 1

Fluid category 1: Wholesome water supplied by a water undertaker and complying with the requirements of regulations made under Section 67 of the Water Industry Act 1991.
Example: Water supplied directly from a water undertaker's main

(b) Determination of fluid category 2

Fluid category 2: Water in fluid category 1 whose aesthetic quality is impaired owing to: (a) a change in its temperature; or   (b) the presence of substances or organisms causing a change in its taste, odour or appearance,   including water in a hot water distribution system.
Examples: Mixing of hot and cold water supplies   Domestic softening plant (common salt regeneration)   Drink vending machine in which no ingredients or carbon dioxide are injected into the supply or distributing inlet pipe   Fire sprinkler systems (without anti-freeze)   Ice-making machines   Water-cooled air conditioning units (without additives)

Continued

Table 6.5   *Continued*

(c) Determination of fluid category 3

Fluid category 3:	Fluid which represents a slight health hazard because of the concentration of substances of low toxicity, including any fluid which contains: (a) ethylene glycol, copper sulphate solution or similar chemical additives; or (b) sodium hypochlorite (chloros and common disinfectants).
Examples:	Water in primary circuits and heating systems (with or without additives) in a house Domestic washbasins, baths and showers Domestic clothes and dishwashing machines Home dialysing machines Drink vending machines in which ingredients or carbon dioxide are injected Commercial softening plant (common salt regeneration only) Domestic hand-held hoses with flow controlled spray or shut-off control Hand-held fertilizer sprays for use in domestic gardens Domestic or commercial irrigation systems, without insecticide or fertilizer additives, and with sprinkler heads not less than 150 mm above ground level

(d) Determination of fluid category 4

Fluid category 4:	Fluid which represents a significant health hazard due to the concentration of toxic substances, including any fluid which contains: (a) chemical, carcinogenic substances or pesticides (including insecticides and herbicides); or (b) environmental organisms of potential health significance.
Examples:	*General* Primary circuits and central heating circuits in other than a house Fire sprinkler systems using anti-freeze solutions *House gardens* Mini-irrigation systems without fertilizer or insecticide application; such as pop-up sprinklers or permeable hoses *Food processing* Food preparation Dairies Bottle-washing apparatus *Catering* Commercial dishwashing machines Bottle-washing apparatus Refrigerating equipment *Industrial and commercial installations* Dyeing equipment Industrial disinfection equipment Printing and photographic equipment Car washing and degreasing plants Commercial clothes washing plants Brewery and distillation plant Water treatment or softeners using other than salt Pressurised fire-fighting systems

Continued

Table 6.5   *Continued*

(e) Determination of fluid category 5

Fluid category 5:   Fluid representing a serious health hazard because of the concentration of pathogenic organisms, radioactive or very toxic substances, including any fluid which contains:
        (a)   faecal matter or other human waste; or
        (b)   butchery or other animal waste; or
        (c)   pathogens from any other source.

Examples: *General* Industrial cisterns Non-domestic hose union taps Sinks, urinals, WC pans and bidets Permeable pipes in other than domestic gardens, laid at or below ground level, with or without additives Grey water recycling systems *Medical* Any medical or dental equipment with submerged inlets Laboratories Bedpan washers Mortuary and embalming equipment Hospital dialysis machines Commercial clothes washing plant in health care premises Non-domestic sinks, baths, washbasins and other appliances *Food processing* Butchery and meat trades Slaughterhouse equipment Vegetable washing *Catering* Dishwashing machines in health care premises Vegetable washing *Industrial and commercial installations* Industrial and commercial plant, etc. Mobile plant, tankers and gulley emptiers Laboratories Sewage treatment and sewage cleansing Drain cleaning plant Water storage for agricultural purposes Water storage for fire-fighting purposes *Commercial agricultural* Commercial irrigation outlets below or at ground level and/or permeable pipes, with or without chemical additives Insecticide or fertilizer applications Commercial hydroponic systems

Note   The list of examples shown above for each fluid category is not exhaustive.

6.6   Wholesite and zone backflow protection

This is used on a supply pipe or distributing pipe where there is an increased risk of backflow because of the following:

- exceptionally heavy use;
- presence of a hazardous substance;
- possibility of internal backflow in buildings of multiple occupation, e.g. flats or tall buildings.

Wholesite and zone backflow protection should be provided in addition to those prevention devices installed at points of use and is required when:

- supply or distributing pipes convey water to two or more separately occupied premises:
- premises are required to provide a storage cistern capacity for 24 hours or more of normal use or premises receiving intermittent supply.

Acceptable arrangements for wholesite or zone backflow protection are shown in figure 6.23.

(a) Wholesite or 3 one backflow protection on supply pipes to separate premises

Single or double check valve depending on level of risks.

No need for secondary protection at lowest level.

No part of the branch pipe to be higher than its connection to the common supply pipe.

Connection of branch pipes to be at least 300 mm above the overflowing level of highest appliance served.

Wholesite protection will protect premises at one level from those at another level, e.g. flats or shops with flat over.

Zone protection will protect one part of a building from another part, e.g. industrial or commercial premises.

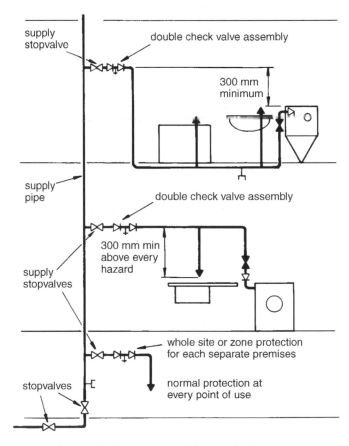

Figure 6.23   Acceptable arrangements for wholesite or zone backflow protection

continued

(b) Wholesite or zone backflow protection on distributing pipes

No part of the branch pipe to be higher than its connection to the common distributing pipe.

Single or double check valve depending on level of risk.

No need for secondary protection at lowest level.

Connection of branch pipe to be at least 300 mm above the overflowing level of highest appliance served.

Wholesite protection will protect premises at one level from those at another level.

Zone protection will protect one part of a building from another part.

(c) Wholesite or zone backflow protection on distributing pipes using a vent pipe

Vent will admit air to prevent backflow.

No need for secondary protection at lowest level.

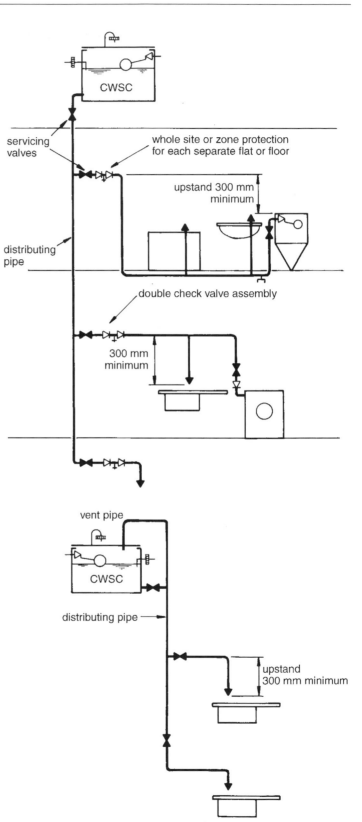

Figure 6.23   **Acceptable arrangements for wholesite or zone backflow protection**
continued

6.7   Application of backflow prevention devices

In general, backflow can be prevented by good system design combined with the use of backflow devices or arrangements chosen to suit the category of risk for which they are designed.

Water Regulations advise that where possible backflow protection should be achieved without the use of mechanical devices. For example, point-of-use protection may include the 'tap gap' or 'air gap' as seen in table 6.4 and figure 6.7. In the case of cistern-fed appliances, zone protection may be provided in the form of permanently vented distributing pipes. In many high-risk situations an air gap is the only device permitted.

Backflow prevention arrangements and devices should be positioned and installed so they are readily accessible for inspection, repair and renewal. It is important that all backflow devices are regularly checked and maintained in good working condition. This applies particularly to mechanical devices.

Bidets and WCs adapted as bidets

Bidets are an obvious contamination risk that come within fluid category 5 of backflow risk. There are two types of bidet: the over-rim type, and the ascending spray or submerged inlet type. Each is protected differently. The former may be used with its taps under mains pressure, the latter is required to be supplied from a storage cistern.

Over-rim types should be arranged with type AUK2 air gaps in accordance with table 6.4 between the tap outlets and the spill-over level of the appliance (see figure 6.24). It is an offence to attach a hand-held flexible spray or similar fittings to taps or bidets where connection is made directly from the supply pipe.

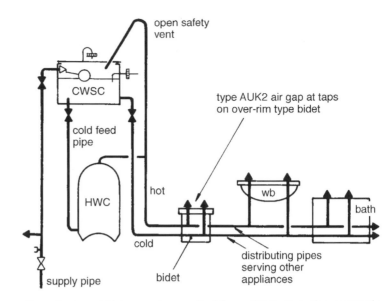

Over-rim type bidets may be supplied from distributing pipe or supply pipe provided type AUK2 air gap is maintained.

Figure 6.24 **Over-rim type bidet**

Ascending spray type bidets, and those having hand-held sprays attached, present a more serious risk than over-rim types and are not permitted to be connected directly from a supply pipe. When making connections to these the following points should be borne in mind:

- They should be supplied with hot and cold water via a break cistern.
- Both hot and cold connections should be supplied from independent, dedicated distributing pipes that do not supply any other appliance; except in the following cases:
  (a) a common cold distributing pipe that serves WC or urinal flushing cistern only in addition to the bidet;
  (b) where the bidet is the lowest appliance in the premises and there is no likelihood of other fittings being fitted at a lower position at a later date and there is no spray attachment fitted, e.g. in a bungalow with all rooms at one level or in a house with a bidet at the lowest floor level.
- The branch pipe connection to the bidet should be at least 300 mm above the spill-over level of the bidet or 300 mm above the highest point that a spray attachment might reach.

Figure 6.25 shows typical acceptable arrangements.

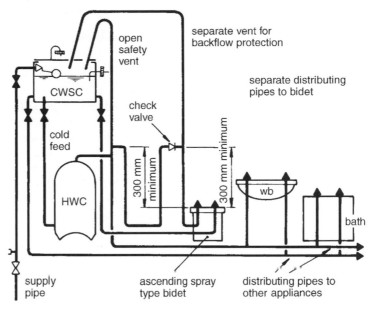

Before fitting a bidet with an ascending spray or flexible hose, the water undertaker MUST be notified (Regulation 5)

(a) Bidet supplied from separate dedicated distributing pipes

Precautions required to prevent backflow from bidet through cylinder to appliances:

○ vent to atmosphere;
○ check valve fitted downstream of vent;
○ 300 mm minimum upstand.

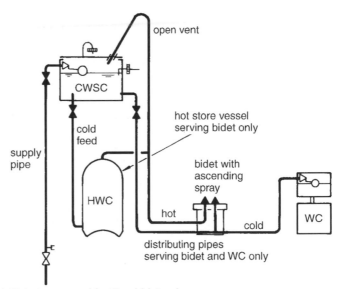

(b) Hot store vessel feeding bidet only

Connection between WC and bidet permitted because this creates no additional hazard.

No special precautions required where hot store vessel feeds the bidet only.

Figure 6.25   Ascending spray type bidet

WCs and urinals

WC pans and urinals are within fluid risk category 5 and present a serious risk of backflow whatever their situation, whether in a house or in industrial or commercial premises. There are two suitable types of backflow device for these:

(1) an **interposed cistern (type AUK1)**, which means a siphonic or non-siphonic flushing cistern, may be used in any type of premises; or

(2) a **pipe interrupter with permanent atmospheric vent (type DC)**, fitted to the outlet of a manually operated pressure flushing valve, may be connected to a supply pipe or distributing pipe (but not in a house). There should be no other obstruction between the outlet of the pipe interrupter and the flush pipe connection to the appliance.

The pipe interrupter should be positioned so that its vent outlet is at least:

- 300 mm above the overflowing level of the WC pan, or
- 150 mm above the urinal bowl being served.

Taps and shower outlets to sanitary appliances

All single outlet taps, combination taps and fixed shower heads should discharge above the appliance, terminating with an air gap as shown in figure 6.7. In domestic premises a tap gap (AUK2) should be used, whilst all other premises should maintain the more stringent AUK3 air gap.

Sinks in both domestic and non-domestic situations are considered to be a fluid category 5 risk and as such the minimum protection is the type AUK3 air gap. However, this is not generally a problem as sinks need additional space for access to work and for the filling of buckets and other appliances (see figure 2.1). Where appliances such as baths and wash basins in domestic premises have submerged inlets they are considered to be a category 3 risk and both hot and cold inlets should be supplied through type EC or type ED double check valves. Appliances for non-domestic use will have a higher category of risk and will need backflow protection to suit the risk. For instance hospitals are a category 5 risk.

Washing machines and dishwashers

Household machines have backflow protection to fluid category 3 built in during manufacture. Before installing an appliance of this kind reference should be made to the *Water Materials and Fittings Directory* in which they will be listed if they are approved under the Water Regulations Advisory Scheme. Commercial machines such as those used in laundromats or similar premises are a category 4 risk, whilst clothes washing plant or equipment in health care establishments are fluid risk category 5.

Drinking water fountains

These should be designed so that there is a minimum 25 mm air gap between the water delivery nozzle and the spill-over level of the bowl.

Additionally, the nozzle should be screened or shrouded to prevent mouth contact.

Outside taps and garden supplies

Backflow protection for hose taps depends on the level of risk for the individual application. Any backflow prevention device used should be fitted inside a building where it will not be subjected to frost damage.

- Hosepipes held in the hand for garden and others uses should be fitted with a *self-closing mechanism* at the hose outlet.
- In a house situation, any garden tap to which a hose connection can be made should be fitted with a *double check valve* positioned inside a building. This will give adequate protection if the hose is held in the hand or where the hose outlet is fixed and provides a permanent air gap (see figure 6.26a).
- Mini-irrigation systems and porous hoses used in a house garden require a *double check valve* as minimum protection and additionally a *pipe interrupter with moving element (type DB)* fitted at the connection of the hose and at least 300 mm above the highest water outlet in the system (see figures 6.26b and 6.26c).
- The *double check valve* is also considered sufficient protection for hand-held hoses used for spraying fertilizers or domestic detergents in household garden situations.
- For more severe risks, such as the application of insecticides, that are considered to be a fluid risk category 5 even in domestic premises, higher levels of backflow control are needed.

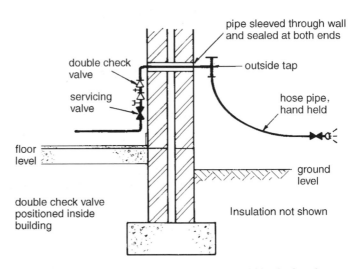

(a) Hose union tap, and tap with hose pipe held in the hand

Figure 6.26   **Backflow protection to external taps in houses**

continued

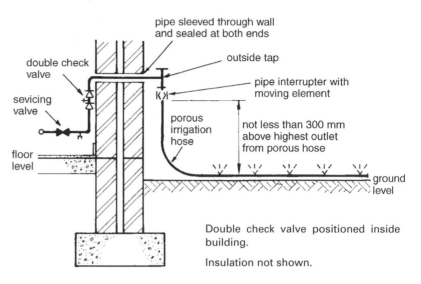

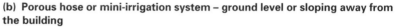

(b) Porous hose or mini-irrigation system – ground level or sloping away from the building

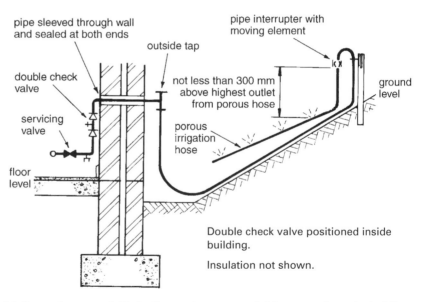

(c) Porous hose or mini-irrigation system – ground rising away from the building

Figure 6.26 **Backflow protection to external taps in houses**
continued

Existing garden taps

In a house situation a hose union tap fitted before the Water Supply (Water Fittings) Regulations 1999 came into force (1 July 1999) may be fitted in one of three ways. Either:

(1) the existing hose union tap should have a *double check valve* installed inside a building; or
(2) the tap should be replaced by one that incorporates a *double check valve arrangement (type HUK1)*; or
(3) a *hose union backflow preventer (type HA)*, or a *double check valve*, should be fitted to the outlet of the tap.

Outside taps and systems in commercial premises

Taps used for non-domestic applications generally present a higher risk than those in domestic premises. Backflow protection should be provided to suit the level of risk and the application, e.g. commercial, horticultural or industrial applications. For example, soil watering and irrigation systems such as permeable hoses are considered to be a fluid risk category 5 and should be supplied only through one of the following backflow prevention devices:

- Type AA air gap with unrestricted discharge;
- Type AB air gap with weir overflow;
- Type AUK1 air gap with interposed cistern.

There may also be a need to provide additional zone protection.

Backflow protection in industrial/commercial premises

There are many industrial and trade premises where risk of contamination may be present, and where backflow protection may be required in addition to any zone backflow protection that is provided. Examples of some of these are shown in the following pages, and include improvements to backflow protection in existing fire sprinkler systems.

Figure 6.27 shows an example of where a property is supplied with water for both domestic and non-domestic purposes.

Spill-over level of appliances served must be at least 300 mm below the invert level of the warning/overflow pipe, and at least 15 mm below the base of the cistern.

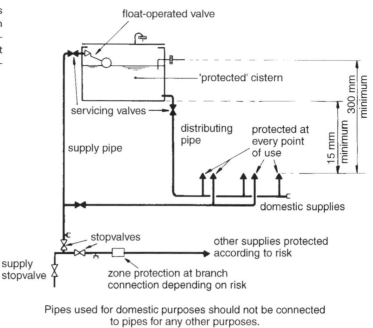

Pipes used for domestic purposes should not be connected to pipes for any other purposes.

Pipes used for domestic purposes should not be connected to pipes for any other purposes.

Figure 6.27   **Supplies for domestic and other purposes**

Figure 6.28 shows examples of the prevention of contamination by backflow or cross connection within industrial, commercial, trade, research, educational, medical and similar establishments, in addition to any zone backflow protection that might be required.

In premises where drinking water and non-drinking water supplies are made available, they should be clearly identified as 'drinking water', 'non-drinking water' or 'fire-fighting water', as appropriate. Hoses intended for drinking water should be used only for that purpose and should be marked 'not for cleaning purposes'.

Water regulations require that 'any fluid that is not wholesome water shall be clearly identified so as to be easily distinguished from any supply pipe or distributing pipe'.

Diagram shows type AA air gap.

Could also use type AB or type AD air gap.

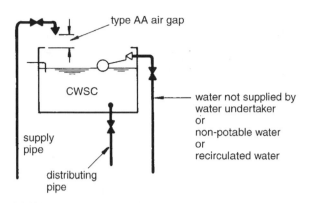

type AA air gap

CWSC

water not supplied by water undertaker
or
non-potable water
or
recirculated water

supply pipe

distributing pipe

(a) Water supplied by undertaker and non-potable water from other sources

Diagram shows type AA air gap.

Could also use type AB or AD air gap

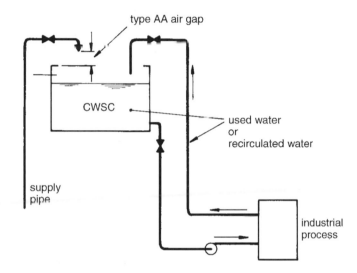

type AA air gap

CWSC

used water
or
recirculated water

supply pipe

industrial process

(b) Re-used or recirculated water

Diagram shows type AB air gap.

Could also use type AA or type AD air gap.

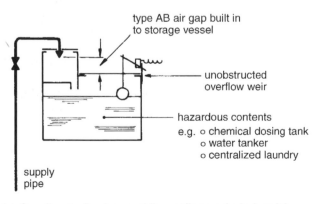

type AB air gap built in to storage vessel

unobstructed overflow weir

hazardous contents
e.g. o chemical dosing tank
o water tanker
o centralized laundry

supply pipe

(c) Supplies to fixed or mobile appliances in industrial processes

Figure 6.28   Backflow protection in industrial and commercial installations

Figure 6.29 shows the prevention of contamination by backflow or cross connection within agricultural and similar establishments (in addition to any wholesite or zone backflow protection).

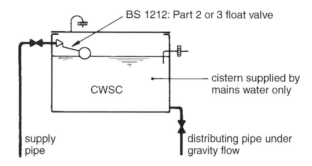

(a) Supplies from storage

Spill-over level of appliances served must be at least 300 mm below the invert level of the warning/overflow pipe, and at least 15 mm below the cistern base.

Where pump is fitted to distributing pipe, cistern must be fitted with type AA, AB or AD air gap arrangement.

Supplies to static or mobile appliances, e.g. crop sprayer, to be connected via type AA or AB air gap or properly connected break cistern serving no other appliance.

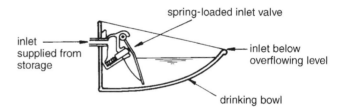

(b) Supply to animal drinking bowl from storage

Drinking bowl without adequate air gap or shrouded inlet should be supplied from storage using an interposed cistern with a dedicated distributing pipe that serves only similar fluid category 5 (high-risk) appliances.

If inlet has type AA air gap and is shrouded from mouth contact, the drinking bowl could be supplied direct from the supply pipe or distributing pipe.

Further backflow protection may be needed if other appliances are supplied from same source.

Figure 6.29   Backflow protection in agricultural and horticultural installations

Continued

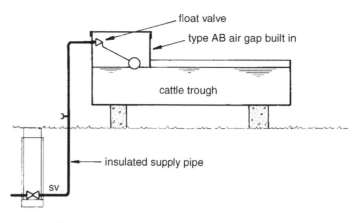

(c) Cattle trough connected to supply pipe

Figure 6.29 **Backflow protection in agricultural and horticultural installations**
continued

Fire protection systems

The scope of this book does not permit detailed discussion of fire protection systems, which are subject to Rules of the Loss Prevention Council and the requirements of the local water undertakers who should be consulted before commencing any installation

Sprinkler systems are commonly installed, particularly in high-risk situations. They are fitted for emergency fire protection only and should be used for no other purpose. Water in them may become stagnant and create contamination risks, particularly where substantial volumes are stored in ground level or elevated cisterns.

Wet spinkler systems without additives, first-aid fire hose reels and hydrant landing valves are fluid risk category 2 and require the minimum protection of a *single check valve* only.

Wet sprinkler systems with additives, and systems containing hydro-pneumatic pressure vessels, are considered to be in fluid risk category 4, and will require either *a verifiable backflow preventer (RPZ valve)* or an air gap, type AA, AB or AD.

Where a common supply pipe serves a fire protection system and a supply pipe for drinking water and domestic purposes, the fire supply should be connected immediately on entry to the building and appropriate backflow protection should be fitted close to the point of connection.

Figure 6.30 shows backflow protection in fire sprinkler systems.

Wet sprinkler systems without additions are fluid risk category 2.

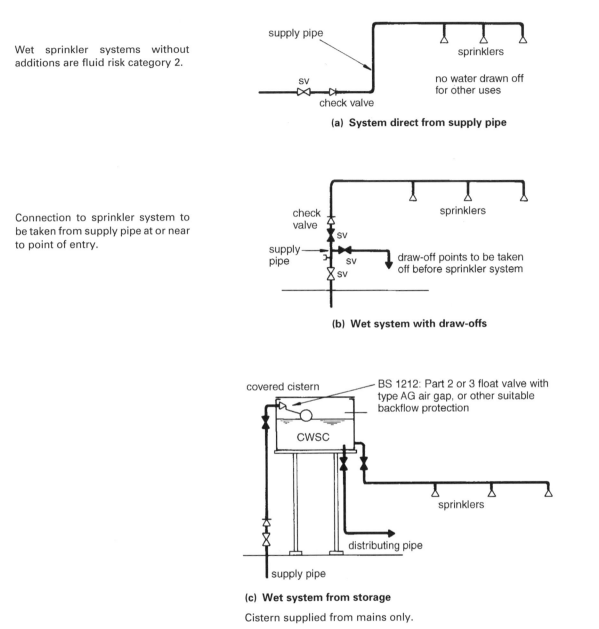

supply pipe

sprinklers

sv

check valve

no water drawn off for other uses

(a) System direct from supply pipe

Connection to sprinkler system to be taken from supply pipe at or near to point of entry.

check valve

sprinklers

sv

supply pipe

sv

sv

draw-off points to be taken off before sprinkler system

(b) Wet system with draw-offs

covered cistern

BS 1212: Part 2 or 3 float valve with type AG air gap, or other suitable backflow protection

CWSC

sprinklers

distributing pipe

supply pipe

(c) Wet system from storage

Cistern supplied from mains only.

Water for other purposes to be drawn separately from storage cistern

Separate cistern for fire purposes preferred.

An uncovered cistern must supply the sprinklers only.

Figure 6.30   Backflow protection in fire sprinkler systems

continued

Type AA air gap shown. Could use
type AB air gap.

Type AA air gap shown. Could use
type AB air gap.

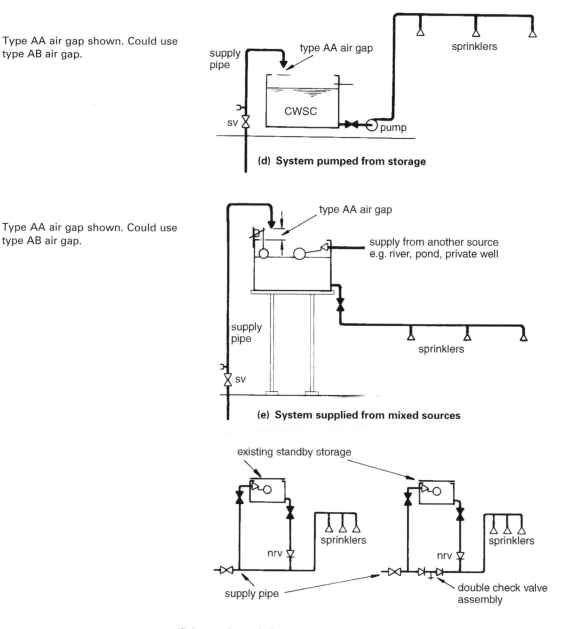

(d) System pumped from storage

(e) System supplied from mixed sources

(f) Improving existing systems

An alternative to the double check valve arrangement would be a pressure
principle backflow preventer. The best solution, provided the effectiveness
of the sprinkler system is not reduced, would be to remove the connection
between the supply pipe and the sprinkler distributing pipe.

This installation only applies to the improvement of protection to existing
buildings. It will not be permitted in new installations.

Figure 6.30   **Backflow protection in fire sprinkler systems**
continued

Chapter 7
Frost precautions and maintenance of water temperature

Water temperatures in systems will vary from time to time depending on a number of factors, for example ambient temperature of the surrounding air, temperature of the incoming water supply, provision of insulation and heat to the building and its pipework systems, much of which, in turn, will be affected by weather conditions and the time of the year.

Temperature changes in systems are inevitable and must be accepted provided they do not become too extreme. In extreme cases protective measures may need to be taken.

Water installations should be protected against the effects of:

- frost and the formation of ice in pipes and fittings;
- loss of heat from hot water pipes, fittings and storage vessels;
- heat gains in cold water pipes, fittings and storage vessels;
- condensation on the surface of pipes.

Heat losses from hot water cylinders, pipes and fittings can cause serious waste of heat energy. Pipes and fittings should be insulated to prevent this. In the case of cylinders and tanks a lagging jacket should be fitted, or preferably hot water storage vessels that are factory-insulated should be used.

Heat gains in cold water pipes can lead to an increase in bacterial activity and subsequently to a deterioration in water quality. This can be offset by using insulation to the standards shown later in table 7.1. Where hot water pipes are run in close proximity to each other, the cold pipe should be placed below the hot one to avoid the effect of rising heat.

Condensation can occur where cold pipes pass through areas of relatively high humidity. The moisture formed may lead to some corrosion of metal pipes or in extreme cases, damage to the building structure.

7.1 Protection from frost

When water loses heat and its temperature drops below freezing point it turns to ice. Upon freezing its volume increases by approximately 10%. This results in:

- damage to pipework and fittings due to the greater volume of ice compared to water;
- the risk of explosion should hot water apparatus be put into operation when part of the system is blocked with ice.

The temperature of water supplied is quite low. In winter, just a small reduction in temperature will cause freezing.

The best precaution against freezing of water services in buildings is the obvious one of keeping the inside of the building continuously warm by the provision and maintenance of adequate heating. This will be easier to achieve and more economical if the building has been designed and constructed so as to minimize the loss of heat through its structure. When the whole building is not heated, or where heating is only intermittent, the localized heating of water pipes and fittings or the heating of their immediate surroundings may suffice. Localized heating, such as trace heating, in conjunction with a frost thermostat, should be used only in addition to other forms of frost protection, or where those are unsuitable.

Location of pipes

Pipes should be located so as to minimize the risk of frost damage, avoiding areas that are difficult to keep warm, especially the following:

- unheated parts of the building, e.g. roof spaces, cellars and underfloor spaces, garages or outhouses;
- areas where draughts could occur, e.g. near windows, air bricks, ventilators, external doors or under suspended floors;
- cold surfaces, e.g. chases and ducts in outside walls and places where pipes come into direct contact with outside walls;
- any exterior position above ground.

If it is not possible to avoid these locations then protection should be provided.

Although some plastic pipes and cisterns are flexible and not easily fractured by ice formation, such formation stops the supply of water. Therefore precautions are still necessary.

7.2   Protection of pipes and fittings

Pipes underground

These should be laid at least 750 mm below the ground surface bearing in mind any expected changes in ground levels (see figure 7.1). This minimum requirement will be sufficient in most cases, but where more severe weather can be expected, then greater depth of cover can be provided up to a maximum of 1350 mm.

The insulating value of the soil will be affected by its nature and water-retaining capacity as well as the degree of exposure of the site. Therefore extra depth of cover may be required to suit local conditions, see figure 7.2.

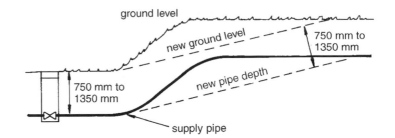

Where ground is to be levelled, ensure pipe is maintained at full depth of cover. This is important, especially on new installations, where ground is often levelled after services are laid.

Minimum depth of cover must be maintained over whole length of pipe.

Figure 7.1 **Pipes in uneven ground**

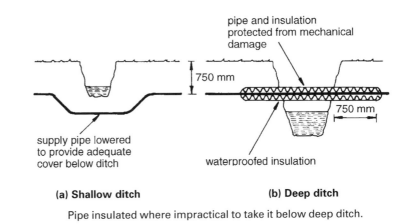

(a) Shallow ditch **(b) Deep ditch**

Pipe insulated where impractical to take it below deep ditch.

Figure 7.2 **Pipes passing under or through ditches**

Stopvalves below ground

Underground stopvalves should not be brought up to a higher level merely for ease of access, but should remain at the same depth as the pipe.

Pipes entering buildings

Pipes should enter buildings at the same depth as laid (see figures 7.3 and 7.4). Where it is not possible to maintain a minimum depth of 750 mm, pipes should be insulated. See also figure 10.3.

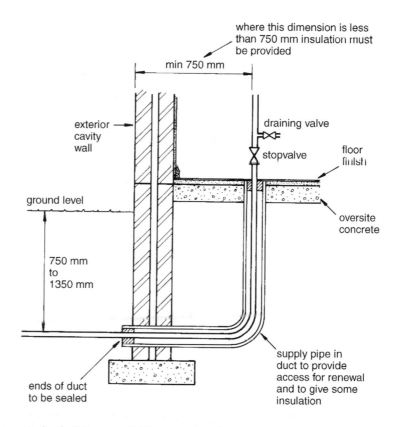

Figure 7.3   **Pipes entering buildings – solid floor construction**

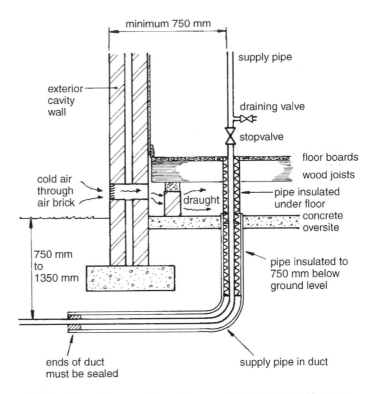

minimum 750 mm

supply pipe

exterior cavity wall

draining valve

stopvalve

floor boards

wood joists

cold air through air brick

draught

pipe insulated under floor

concrete oversite

750 mm to 1350 mm

pipe insulated to 750 mm below ground level

ends of duct must be sealed

supply pipe in duct

Shallow foundations as shown here would not be normal in new buildings, but where this situation does occur the pipe beneath the foundation should be protected from settlement by the building.

Figure 7.4   Pipes entering buildings – suspended timber floor

Pipes and fittings outside buildings

Where possible these should be laid underground. Where this is not possible sufficient protection against the effects of frost should be provided. See figures 7.5 and 7.6.

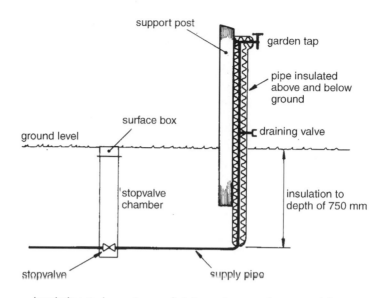

Insulation to be waterproofed throughout and protected from damage.

Note   Backflow protection device needed if hose pipe is to be connected.

Figure 7.5   Pipe rising from below ground to garden stand-pipe

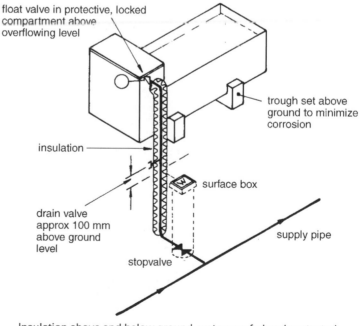

Insulation above and below ground, waterproofed and protected from damage by animals.

Figure 7.6   Pipe rising from below ground to cattle trough

Pipes and cisterns in or above roof spaces

When considering the insulation of roof spaces and pipes and components within roof spaces (see figure 7.7), full advantage should be taken of any heat rising from rooms. Roof spaces are notoriously cold and draughty places and the cause of many frozen pipes in the UK. This has not been helped by Building Regulations which require ventilation of roof spaces to prevent condensation. Consideration should also be given to the following:

- ceiling insulation should be omitted below cisterns;
- insulation should be provided all over any cistern and include any pipes rising to connect with it;
- pipes within roof spaces, including warning and overflow pipes, should wherever possible be fixed below any ceiling insulation;
- any pipes other than those situated as above should be adequately insulated.

If a cistern is sited above the roof of a building, it must be protected by installing it, together with its inlet and outlet pipes, in an insulated enclosure provided with either its own means of heating, or opening into some heated part of the building itself.

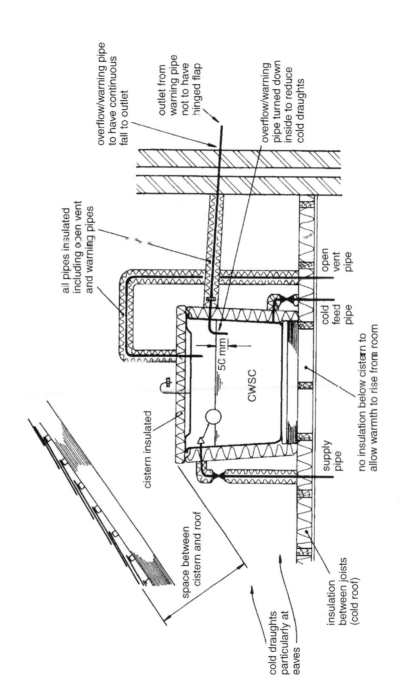

overflow/warning pipe
to have continuous
fall to outlet

outlet from
warning pipe
not to have
hinged flap

overflow/warning
pipe turned down
inside to reduce
cold draughts

all pipes insulated
including open vent
and warning pipes

open vent
pipe

cold
feed
pipe

cistern insulated

50 mm

CWSC

no insulation below cistern to
allow warmth to rise from room

supply
pipe

space between
cistern and roof

insulation
between joists
(cold roof)

cold draughts
particularly at
eaves

Figure 7.7   Insulation of pipes and cisterns in roof space

7.3   Draining facilities

It should be possible to drain down all parts of a hot or cold water system including its pipes, fittings, components and appliances, so that pipes not in use can be emptied in cold weather. An empty pipe cannot suffer frost damage. For pipes to be adequately drained, allowance must be made for the entry of air. Usually this will be done through taps and vent pipes but in some special cases air inlet valves will be needed.

To assist draining:

- pipes should be laid to slight and continuous falls to draining valves at low points;
- draining taps should have means for connection of a hose pipe;
- cisterns, cylinders and tanks should be fitted with draining taps unless they can be drained through pipes leading to a draining tap elsewhere.

Draining taps must be of the screw-down pattern having a removable key and must comply with BS 2879 (see figure 7.8).

BS 6700 recommends that all pipes and fittings be drainable. However, it is the author's view that there are some exceptions to this rule. Draining taps should not be fitted within sealed ducts or below floors or below ground where they may be inaccessible (see figure 7.9). More importantly, they should not be fitted where they might become submerged and create a contamination risk (see figure 7.10).

Where a building is divided into parts, the pipework should be arranged so that each part can be isolated and drained without affecting the other parts.

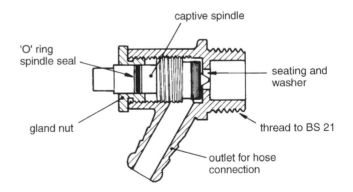

Figure 7.8   **Drain cock to BS 2879**

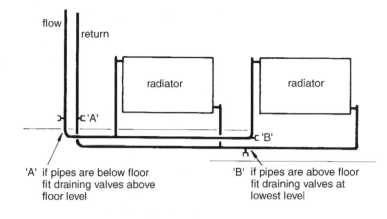

'A' if pipes are below floor
fit draining valves above
floor level

'B' if pipes are above floor
fit draining valves at
lowest level

For pipe drops below floor level where the floor is likely to be covered by carpets and furniture it may be better to fit the draining valves above, but near the floor (alternative A).

Alternative B is more suitable where pipes do not drop below floor level.

Figure 7.9   Draining arrangements – pipe drops below floors

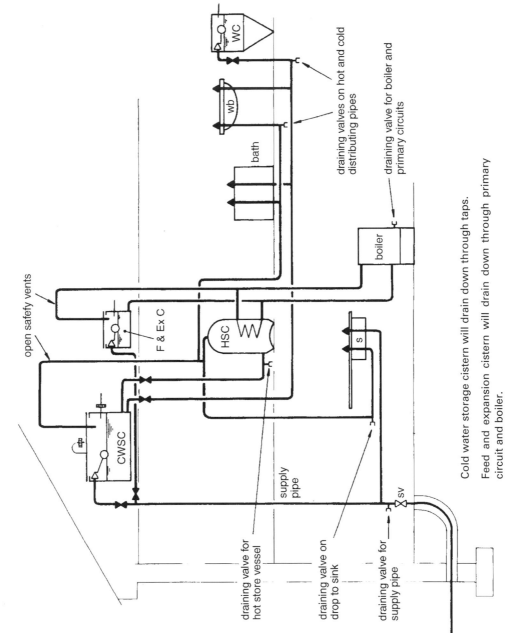

Cold water storage cistern will drain down through taps.

Feed and expansion cistern will drain down through primary circuit and boiler.

Figure 7.10   Hot and cold water system showing draining valve positions

7.4   Insulation

The thickness of insulation used to protect an installation is dependent on the following factors:

- the type of insulation used and its thermal conductivity value;
- the reason for the insulation, whether to protect against frost damage, heat gains, heat losses, or condensation; or
- whether the pipes, fittings and components are in heated or unheated premises, or whether they are indoors or outdoors.

BS 5422 shows insulation tables to suit a wide range of contingencies for various types of building usage. Tables 6 and 7 of BS 6700 give calculated figures, based on BS 5422, for the protection of copper and steel pipes in both domestic and commercial premises, respectively. Unfortunately, these do not have a practical application.

Guidance to Schedule 2 of Water Regulations discusses the background to the criteria used in BS 5422 for calculating insulation, and considers two conditions, normal and extreme, where differing thicknesses of insulation might apply.

Normal conditions would include water fittings installed within buildings in areas that are not subjected to draughts from outside the building. Examples given are cloakrooms, store rooms, utility rooms, roof spaces where pipes are below the ceiling insulation, and unheated parts of otherwise heated commercial buildings.

Extreme conditions include water fittings installed inside unheated or marginally heated buildings, or below suspended floors, cold roof spaces or other areas where draughty conditions are likely. Exposed conditions will obviously include water fittings installed outside of buildings.

Guidance to Water Regulations discusses the need for insulation against frost damage in normal and extreme conditions but only gives recommended thicknesses of insulation for copper pipes complying with BS EN 1057 in normal exposures. This table is reproduced here as table 7.1. The Guidance goes on to suggest that an increased insulation value should be applied to extreme conditions, and that the advice of insulation manufacturers should be sought.

Building Regulations require that all hot water pipes, fittings and storage vessels be insulated. Tables for the economic insulation of hot water pipes are shown in BS 5422 and were reproduced earlier in tables 3.3 and 3.4 and examples of insulating material are given in table 7.2. (Energy conservation is discussed later in section 8.2.)

Space should be allowed around pipes for the insulation which should be continuous over pipes and fittings with space left only for the operation of valves.

Where necessary, insulating material must be resistant to, or should be protected from, mechanical damage, rain, moist atmosphere, subsoil water and vermin. If the insulation is not water-resistant or in vulnerable

Table 7.1 **Recommended minimum commercial thicknesses of thermal insulation for copper water pipes of minimum wall thickness complying with BS EN 1057 in normal conditions of exposure**

External diameter of pipe (in mm)	Thermal conductivity of insulation material at 0°C in W/(m.K)				
	0.02	0.025	0.03	0.035	0.04
	Thickness of insulation (in mm)				
15	20 (20)	30 (30)	25* (45)	25* (70)	32* (91)
22	15 (9)	15 (12)	19 (15)	19 (19)	25 (24)
28	15 (6)	15 (8)	13 (10)	19 (12)	22 (14)
35	15 (4)	15 (6)	9 (7)	9 (8)	13 (10)
42 and over	15 (3)	15 (5)	9 (5)	9 (5)	9 (8)

Notes
(1) Except for 15 mm pipes with thermal conductivities of 0.030, 0.035 and 0.40 W/(m.K), shown with a *, which are limited to 50% ice formation after 9, 8 and 7 hours respectively, the above recommended commercially available minimum thicknesses of insulation should limit ice formation to under 50% after 12 hours for the remainder of the pipe sizes, when based on an air temperature of $-6°C$ and a water temperature of $+7°C$. The minimum calculated insulation thicknesses for 12 hours protection under the above conditions are shown in the appropriate location in brackets.
(2) Commercial thicknesses of insulation with the higher thermal conductivities are generally limited to a minimum of 9 mm. Materials with a lower thermal conductivity, such as rigid phenolic foam, polisocyanurate foam and rigid polyurethane foam are installed by specialist firms and are usually limited to a minimum thickness of about 15 mm.
(3) Normal conditions to frost exposure are considered to be when water fittings are installed inside buildings within the thermal envelope, but within rooms or voids which are not heated for a minimum period of 12 hours each day for the whole of the winter period. Examples could include the following.
(a) unheated cloakrooms, store rooms, utility rooms, etc.
(b) below the ceiling insulation in a roof space.

Table 7.2 **Examples of insulating materials**

Thermal conductivity W/(m.K)	Material
Less than 0.020	Rigid phenolic foam
0.021 to 0.035	Polyurethane foam
0.04 to 0.055	Corkboard
0.055 to 0.07	Exfoliated vermiculite (loose fill)

areas such as in open air or below ground the insulation should be protected with a suitable waterproof material to keep the insulation dry.

Where pipes are fixed to inside surfaces of outside walls there is no need to insulate them against the effects of frost. There is, however, a need to insulate against heat gains in cold pipes, heat losses in hot pipes, and as a means of preventing condensation. It is important that these pipes are not fixed in direct contact with the outside wall, rather, they should be secured with spacer clips or, if saddle clips are used, on to a pipe board (see figure 7.11).

Manufacturers' literature will give details of application and use of insulating materials in various situations.

Note that no amount of insulation can prevent a pipe from freezing. It can, however, slow down the heat loss and delay the effects of frost.

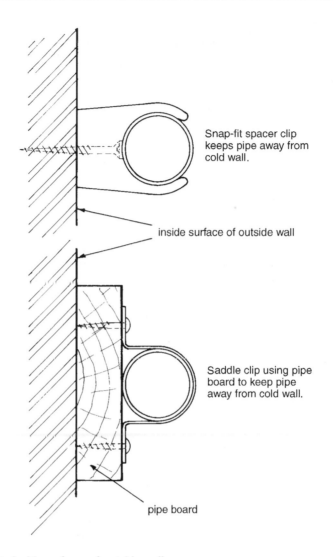

Snap-fit spacer clip keeps pipe away from cold wall.

inside surface of outside wall

Saddle clip using pipe board to keep pipe away from cold wall.

pipe board

Figure 7.11   **Fixings to inside surfaces of outside walls**

Trace heating

Trace heating can be used as a form of local heating to warm pipes and fittings during periods of extreme cold. Trace heating should conform to BS 6351: Part 1 and should be used in addition to any insulation provided and *not* as a substitute. Trace heating should be fitted before the insulation is applied so that the trace heating elements are insulated along with the pipes and fittings.

Local heating

Local heating methods may be used in areas where other forms of protection are impractical or unsuitable, for example in unheated roof spaces or garages. More commonly, local heating is seen to be used in conjuction with a frost thermostat in tank rooms and pump rooms (see figure 2.40).

Chapter 8
Water economy and energy conservation

The 1987 edition of BS 6700 suggested a relationship between water usage and energy consumption, based on the cost of supplying water and disposing of effluent. It also suggested that metering will produce considerable saving in water consumption.

There is a need to conserve water, because as we consume more, sources become increasingly costly and difficult to find. It is, therefore, in the interest of the consumer, and of concern to the plumber, that water is not consumed unduly.

With regard to metering and the cost of water supply, the author only partly agrees with the previous standard's philosophy. Whilst metering will most likely help to reduce the consumption of water, the cost to the consumer must increase because of the difficulties of reading and maintaining meters, and the additional cost of purchasing and installing meters and ancillary materials. Under the present system consumers on higher rateable values and low usage who opt for meters, do receive smaller water bills but, it is feared, at the expense of the majority whose properties are on lower rateable values, or those with large families.

As metering becomes more widespread or mandatory, then the costs of installing, maintaining and reading meters must in the long run cause a general increase in payments for water. At the same time, the energy consumed in manufacturing, installing and maintaining meters and their associated components is likely to exceed the energy savings from reduced water consumption.

Designers of water systems should be constantly aware of the need to keep water usage and energy consumption to a minimum. The following notes have this need in mind.

8.1 Water economy

Water byelaws have always been written in order to prevent or reduce wastage or excessive use of water. Current Water Regulations which came into effect on 1st July 1999 are no exception. Water is costly to produce, and during hot dry spells it is becoming increasingly difficult to maintain a constant supply. BS 6700 was written with water economy in mind and suggests a number of ways in which savings can be made and Water Regulations now take this a step further.

Leakages

Approximately 30% of all the potable water produced in Britain is lost through waste, undue consumption or misuse. Much of this is due to leakages underground from mains and service pipes. Water undertakers carry out waste detection programmes to reduce this loss, but because of the age and condition of many of our underground pipes, it is a continuing problem which can only be solved by regular testing and monitoring of internal and external pipework systems. (See chapter 12 which includes methods for the detection of leakages in metered and unmetered supply pipes.)

Additionally, BS 6700 suggests that:

(1) Warning pipes from cisterns should discharge where they can readily be seen, e.g. over a doorway where the discharge may cause a nuisance.
(2) Ponds and pools should be built so that water loss is kept to a minimum. They should not lose more than 3 mm depth of water per day, after taking rainfall and evaporation into consideration. Ponds and pools must be fitted with an impervious lining or membrane. Any pond or pool installation of more than 10 000 l capacity and designed to be replenished by automatic means, must be notified to the water undertaker before the installation is carried out.

Flushing of WCs and urinals

WC flushing accounts for about 25% of all domestic water used in buildings. Urinals consume water in large quantities if not properly controlled.

The Water Supply (Water Fittings) Regulations 1999 have introduced new requirements for the flushing of WCs and urinals. In brief, these are aimed at bringing us into line with European practices and encouraging new and innovative flushing arrangements with the result that they introduce to this country a number of changes to previous practices.

WC flushing cisterns

To reduce the amount of water used in flushing, Water Regulations now require that sizes of cisterns used for the flushing of WCs be reduced compared to those previously permitted.

WC cisterns are permitted with non-siphonic flushing devices. Typical of these are those used on the continent with valve type actuators using gravity flow rather than the old familiar siphonic arrangement.

Cisterns with siphonic flushing arrangements are still permitted but *all* WC cisterns are required to meet the strict performance criteria set out in the 'Water Regulator's Specification for the performance of WC suites installed after 1st January 2001'. Cisterns, when flushed, must be capable of clearing the contents of the pan effectively using a single flush and flushing volumes for WC cisterns are reduced from 7.5 l to a maximum of 6 l.

The dual-flush cistern is back in fashion. Manufacturers are encouraged to produce dual-flush cisterns that are easier and more positive to use than previous ones. Dual-flush cisterns are required to give a maximum full flush of 6 l and a lesser flush volume of two-thirds that of the full flush, i.e. 4 l.

Also permitted are 'pressure flushing cisterns' which use incoming water pressure to compress air which in turn is used to increase the pressure of water available for flushing the WC pan.

Implementation dates for the new WC flushing arrangements under the Water Regulator's Specification are shown in table 8.1. Figures 8.1–8.3 illustrate the new WC flushing arrangements.

Table 8.1   Maximum permitted capacity of WC cisterns

Type of appliance	Use	Maximum permitted volume (litres)	Dates and restrictions
Single flush and dual flush	Domestic and non-domestic	7.5	Until 1 January 2001 Flushing by cistern with siphonic device only
Single flush and dual flush	Domestic	6	From 1 January 2001 Flushing by cistern only; siphonic and non-siphonic devices permitted
Single flush and dual flush	Non-domestic	6	From 1 January 2001 Flushing by cistern or by pressure flushing valve; siphonic and non-siphonic devices permitted

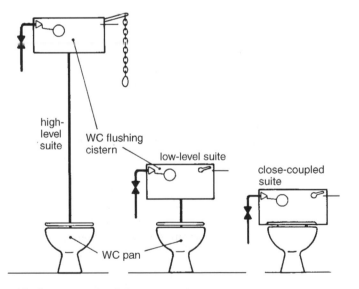

Maximum permitted cistern capacity:

○ single flush 6 l after 1 January 2001
          7.5 l until 1 January 2001

○ dual flush 6 l/4 l after 1 January 2001
          7.5 l/5 l until 1 January 2001

Figure 8.1   Capacity of WC flushing cisterns

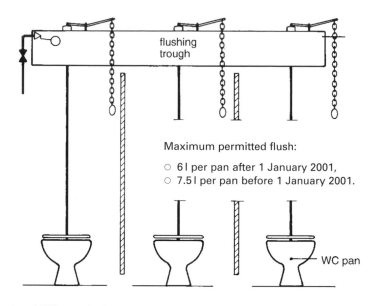

flushing
trough

Maximum permitted flush:

○ 6 l per pan after 1 January 2001,
○ 7.5 l per pan before 1 January 2001.

WC pan

Figure 8.2   Capacity of WC trough cisterns

WC pans are designed to suit the associated cistern to ensure effective cleansing. Care should be taken that the correct cistern or pan is fitted when either is renewed otherwise effective pan clearance may not be achieved.

Pressure flushing valves

As an alternative to the flushing cistern, 'pressure flushing valves' are approved for the flushing of both WCs and urinals. Flushing valves have been used on the continent for many years and have been extensively used in ships where the sea provided more than ample water for flushing. However, flushing valves are *not* permitted for the flushing of WCs in private dwellings, or anywhere that a minimum flow rate of 1.2 l/s cannot be achieved at the appliance. See figures 8.3 and 8.4.

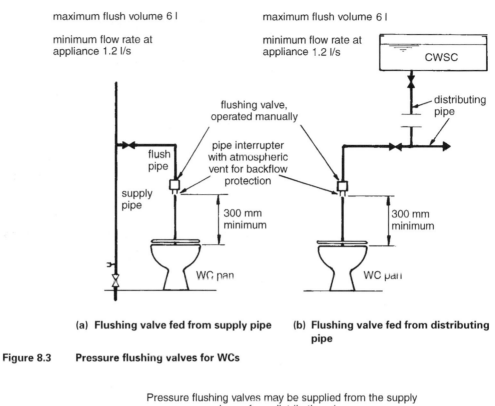

(a) **Flushing valve fed from supply pipe**

(b) **Flushing valve fed from distributing pipe**

Figure 8.3   Pressure flushing valves for WCs

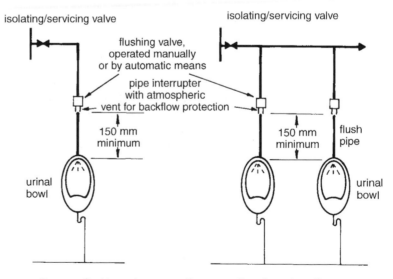

(a) **Pressure flushing valve serving single urinal bowls**

(b) **Pressure flushing valve serving range of urinal bowls**

Figure 8.4   Pressure flushing valves for urinals

Flushing arrangements for urinals

Water Regulations require that automatic flushing cisterns should supply a maximum of 10 l per hour, per bowl, stall, or 700 mm length of slab, delivered not more than three times per hour (see figure 8.5).

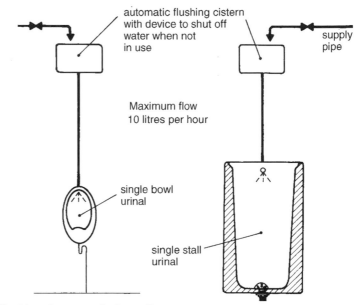

automatic flushing cistern
with device to shut off
water when not
in use

supply
pipe

Maximum flow
10 litres per hour

single bowl
urinal

single stall
urinal

Figure 8.5   Urinal flushing cisterns – single appliances

Additionally, supply pipes to urinal flushing cisterns are required to be fitted with devices that will control the supply during periods when the urinal is not in use. There are two methods of controlling the supply: time control (see figure 8.6), and hydraulic control (see figures 8.7 and 8.8).

Where a time control is used, the pipe supplying the urinal cistern should also be fitted with a lockable control valve.

Alternatively, individual bowls or stalls may be user-operated as required, e.g. by a manual chain pull (see figure 8.9) or push button operation or infra-red sensor or similar. A multiple-style urinal installation is shown in figure 8.10.

Requirements for the maximum flushing volumes of water to be used in the flushing of urinals are shown in table 8.2.

It may not always be feasible to size the automatic flushing cistern for multiple urinals according to the number of bowls, stalls or slabs, or to give the age-old recommended three flushes per hour.

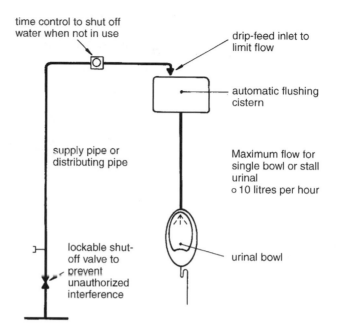

Figure 8.6   Time control to automatic flushing cistern

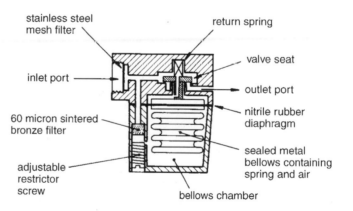

Figure 8.7   Hydraulic flow control device for automatic flushing cistern

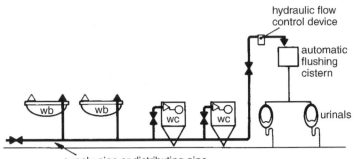

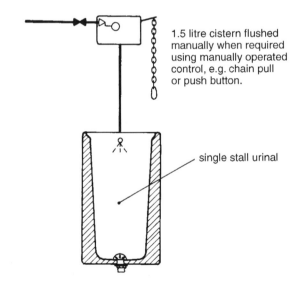

supply pipe or distributing pipe

No use of appliance = no flow through flow control device.

Draw-offs at other appliances will create pressure variations causing the hydraulic flow control to open, thus permitting a small quantity of water to pass into the automatic flushing cistern.

Figure 8.8   Use of hydraulic flow control device

1.5 litre cistern flushed manually when required using manually operated control, e.g. chain pull or push button.

single stall urinal

Figure 8.9   Single urinal with manual flushing control

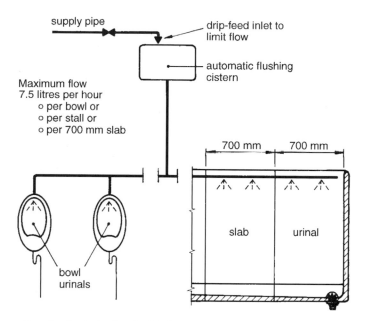

Supply pipe to be fitted with device to shut off water when not in use.

Figure 8.10   **Multiple urinal installation**

Table 8.2   **Maximum permitted volumes of water for flushing urinals**

Appliance	Maximum volume
For a single bowl or stall supplied from an automatic flushing cistern	10 l/hour
For more than one appliance supplied from an automatic flushing cistern	7.5 l/hour per bowl, stall, or per 700 mm width of slab
For individual bowls or stalls, user-operated, and supplied via manual chain pull or button operated from cistern or pressure flushing valve	1.5 l per flush as required

Table 8.3 provides for a variation of flushing arrangements based on the formula:

$$\text{Time interval (min)} = \frac{\text{cistern capacity (l)} \times \text{time in minutes (60)}}{\dfrac{\text{maximum flush requirement}}{7.5\,\text{l per bowl}} \times \dfrac{\text{number of bowls, stalls}}{\text{or 700\,mm width of slab}}}$$

For example, using a 13.5 l cistern to flush five urinal bowls

$$\text{Time interval} = \frac{\text{cistern capacity (13.5 l)} \times \text{time (60)}}{\text{maximum flush (7.5 l per bowl)} \times \text{number of bowls (5)}}$$

$$= \frac{13.5 \times 60}{7.5 \times 5}$$

$$= 21.6\,\text{mm}$$

Table 8.3   **Volumes and flushing intervals for urinals**

Number of bowls, stalls, or per 700 mm of slab	Volume of automatic flushing cistern l				Maximum fill rate l/h
	4.5 l	9 l	13.5 l	18 l	
	Shortest period (minutes) between flushes				
1	27	54	81	108	10
2	18	37	54	72	15
3	12	24	36	48	22.5
4	9	18	27	36	30
5	7.2	14.4	21.6	28.2	37.5
6	6	12	18	24	45

The reduction in quantity of water used should lead to the reduction of limescale build-up in appliances and drains, and consequent reduction in odours.

Water supplies for WC and urinal flushing

For a great many years the water industry resisted the direct connection of WC pans to supply and distributing pipes and strict measures were used, i.e. an interposed cistern (flushing cistern) to guard against the possibility of backflow from these appliances to the main or to other water fittings.

Under Water Regulations, pressure flushing cisterns for WCs and pressure flushing valves for both WCs and urinals may be supplied from a supply pipe or a distributing pipe, providing appropriate backflow prevention devices are in place; the device being a 'pipe interrupter with permanent atmospheric vent'.

Waste plugs

Waste plugs should be fitted to all baths, basins, sinks or similar appliances, except where delivery is less than 3.5l/min and the appliance is designed not to have a plug, e.g. basins with spray taps, and shower trays. Appliances for medical or veterinary purposes are also excepted.

Self-closing taps

Self-closing taps should be of the non-concussive type (see figure 8.11) and be capable of closing against 2.6 times the working pressure. These taps are very effective when new, but tend to fail in the open position after a period in use so they should only be used in buildings where regular maintenance and inspection can be ensured.

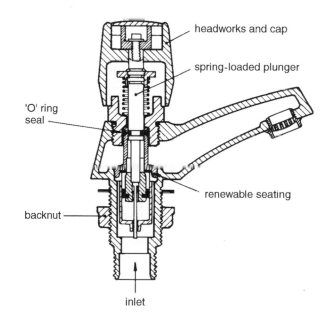

Figure 8.11   Non-concussive self-closing tap

Spray taps

Spray taps can provide savings of up to 50% in both fuel and water. However, they have several disadvantages.

- they should not be used where basins are subject to heavy fouling by grease or dirt;
- they require regular maintenance;
- they are only suitable for hand rinsing;
- the heads may block in hard water areas;

- since self-cleansing velocities of waste water may not be achieved, residues may build up in waste pipes.

Aerators

Aerators may reduce consumption and when compared with spray taps will give improved flow pattern.

Showers

Showers are generally said to use less water than baths. However, the reduced consumption is often offset by more frequent use. Also, with the arrival of pressurized hot water systems, water consumption for showering will increase.

Washing troughs and fountains

Fittings serving these appliances should be capable of discharging to individual units without discharging to other units. A unit means a 600 mm length of straight trough or of the perimeter of a round appliance.

Domestic appliances

Maximum amounts of water used per complete cycle of operations:

Clothes washer without water-using tumbler dryer:	27 l per kg of washload for standard 60°C cotton cycle
Clothes washer with water-using tumbler dryer:	48 l per kg of washload for standard 60°C cotton cycle
Dishwasher:	4.5 l per place setting

Other economy measures

The following measures may also improve water economy:

- protection from mechanical damage and corrosion especially underground (trench preparation and backfill, pipe depth);
- use of approved fittings, components and pipes;
- adequate frost precautions;
- use of spray taps.

8.2   Energy conservation

It is a requirement of Building Regulations that reasonable provision be made for the conservation of fuel and power in buildings. This requirement includes the need for effective controls on hot water and space heating systems, and the use of insulation to prevent undue loss of heat from hot water storage vessels, pipes and fittings and other components.

Hot water storage vessels must be fitted with adequate thermal insulation. Insulating materials should limit heat loss to $90\,W/m^2$ of surface area of the storage vessel. The hot store vessel should be fitted with a thermostat to keep the water at the required temperature, and a time switch that will shut off the heat source when there is no demand. Also energy can be saved by reducing the quantity of water heated, bearing in mind the methods of heating or controlling the heat input to the storage vessel.

Hot store vessels should be of adequate capacity without being oversized. Methods of heating should, where possible, enable a reduced quantity of water to be heated when desired. Examples of this include the use of double element or twin element immersion heaters in electrically heated systems and manually controlled economy valves on gas circulators.

Pipes to and from hot water apparatus and central heating components should be insulated unless they are designed to contribute to space heating. Table 8.4 gives maximum lengths permitted for hot water distributing pipes without insulation. However, it is recommended that all pipes for hot water supply should be lagged.

Maximum rates of heat loss from hot water pipes, when insulated,

Table 8.4 **Maximum permitted rates of energy loss from pipes**

Outside pipe diameter* mm	Maximum energy loss W/m^2
10	676
20	400
30	280
40	220
50 and above	175

* Intermediate pipe diameters may be found by interpolation.

should not exceed those shown in table 8.4. Thickness of insulation for practical purposes should equal that shown in tables 3.3, 3.4 and 7.1.

Trace heating used for the heating of hot water pipes should be of the self-regulating type, and the system should comply with BS 6351.

The use of booster pumps for both hot and cold water systems should be minimized and, in consultation with the water undertaker, consideration should be given to the prudent use of mains water pressure. The energy needed to boost water pressure in a building is about $0.02\,kW/m^3$ per metre of lift. Therefore it is advisable, where mains pressure is insufficient to supply the upper floors of a building, to use mains pressure to the limit of its supply.

Chapter 9
Noise and vibration

Water Regulations require water fittings to be constructed of materials that are resistant to damage from vibration.

Noise is caused by vibration and is generally seen (or heard) to be a nuisance to building occupiers, but vibration associated with pipework noise can also at times cause damage to pipes and fittings leading to leakage within the system.

It is unfortunate that this important aspect of water installations has been omitted from BS 6700, despite its earlier inclusion in the 1987 edition. Because of its importance, the author has decided to continue its inclusion in this book.

In water systems many materials are susceptible to vibration and will transmit or even accentuate noises produced. Most system noises, which many operatives explain away as unavoidable, can be avoided by better design and workmanship.

9.1   Flow noises

Pipework noise becomes significant at water velocities over 3 m/s. It is important therefore that systems are designed to keep water velocities below 3 m/s by increasing the pipe size, as necessary. The causes of some common flow noises are shown in figures 9.1, 9.2 and 9.3.

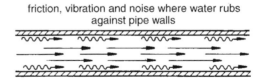

friction, vibration and noise where water rubs against pipe walls

Where velocities are below 3 m/s noise is not significant.

Figure 9.1   Flow noise in pipes

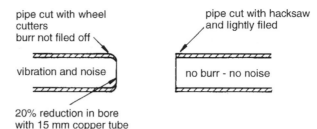

pipe cut with wheel cutters
burr not filed off
vibration and noise
20% reduction in bore with 15 mm copper tube

pipe cut with hacksaw and lightly filed
no burr - no noise

Burrs left on pipe ends after cutting will add to turbulence in pipes and increase noise levels

Figure 9.2   Effect on water flow of burrs on pipe ends

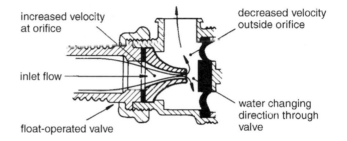

increased velocity at orifice
inlet flow
float-operated valve

decreased velocity outside orifice
water changing direction through valve

Changes in flow direction, and drop in pressure through valve may lead to cavitation which will further increase flow noise. Can be reduced by lowering flow velocity.

Figure 9.3   Noise caused by water flow through narrow apertures

Cavitation

Cavitation can be simply described as wear or erosion of the internal surfaces of pipes and fittings caused by turbulent water flow.

When cavitation occurs, water flow noise increases. Cavitation is not common in pipework as it usually only occurs at water velocities of 7 m/s to 8 m/s in elbow fittings. However, it can occur through reduced pressures at the upper parts of systems which incorporate long pipe drops.

Outlet fittings generally incorporate abrupt changes of direction in water flow, and there is a sudden drop in pressure at the outlet side of the seating of taps and float valves. These conditions are ideal for cavitation, and are the major cause of noise in these fittings. Although figure 9.4 shows a bib tap to BS 1010, this is a problem common to most types of taps and valves.

Cavitation noises can be controlled by reducing the pressure drop across the valve seating. For example, if the inlet pressure is reduced and the tap is opened more fully it will operate more quietly. Similarly, a float valve with a lower inlet pressure and larger orifice will operate with less noise.

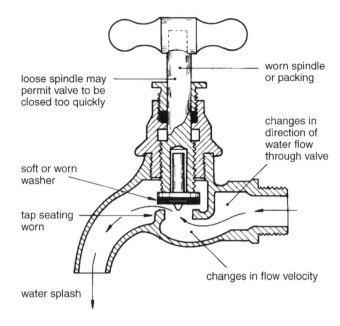

loose spindle may permit valve to be closed too quickly

worn spindle or packing

changes in direction of water flow through valve

soft or worn washer

tap seating worn

changes in flow velocity

water splash

Figure 9.4   **Causes of noise in taps and valves**

Fast flow through narrow apertures

Noises in float valves and stopvalves can often mean that the waterway is becoming blocked with particles carried along in the water (see figure 9.5). To avoid this, stopvalves should always be left in the fully open position so that particles can pass through and out of the system.

In the past, a good way to reduce the problem of blockage, and also reduce general flow noise in float valves, was to fit an equilibrium float valve. The use of the Portsmouth type valve is restricted under the Water Regulations, but there are diaphragm type valves shown in the *Water Fittings and Materials Directory* in sizes from 1″ (G1) upwards. There is also a ceramic disc type listed that provides full flow during fill and is designed to reduce the possibility of water hammer noise.

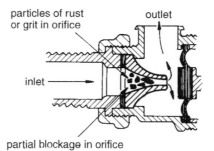

particles of rust or grit in orifice

outlet

inlet

partial blockage in orifice

Any obstruction in a valve which significantly reduces the bore will create noise. Commonly found in:

○ stopvalves that are partially closed;
○ stopvalves that are partially blocked;
○ float valves partially blocked (as illustrated).

It is the mistaken belief of many that to partially close a stopvalve will reduce pressure. In fact, it will create noise and unnecessary wear to the valve.

Figure 9.5   **Noise caused by obstruction in valves**

9.2   Water hammer noise

Valve closure

Sudden valve closure will cause shock waves to be transmitted along pipes with a loud hammering noise. Rapid closure can be prevented by regular maintenance, making sure that packing glands are correctly adjusted and spindles are not loose. Other precautions include the restriction of velocities and avoidance of long straight pipe runs. Limiting water velocities to 3 m/s will not, in itself, reduce water hammer, but will help reduce the magnitude of pressure peaks produced.

Solenoid valves and self-closing taps

These often cause water hammer noise. Use non-concussive types, properly and regularly maintained.

Vibration of the cistern wall

This is very common in copper cisterns (see figure 9.6). It causes the float valve to open and shut in rapid succession causing water hammer, or sometimes a noise similar to that of an electric motor. The cure is to stabilize the cistern wall with a strap around it, and to fix the supply pipe securely to prevent it and the connected float valve from vibrating.

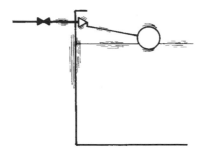

Vibrating cistern wall will move float body and create noise.

Figure 9.6   Vibration of cistern wall

Float valve oscillation

This causes a regular series of loud banging noises which occur when the valve is almost closed. This type of noise can be most disturbing to occupiers of buildings, and could well cause damage to pipes and fittings. There are a number of reasons for it, and BS 6700 (1987 edition) suggested that the commonest cause is the formation of waves on the surface of the cistern water. The author does not agree entirely, and believes the waves are a result of the oscillation and not the cause. However, the problem is there whichever is the case!

For the float valve to operate correctly (see figure 9.7), its closing force (float buoyancy) must overcome the incoming force of the water pressure. If the incoming pressure and the buoyancy force nearly balance, the conditions for oscillation are ideal. A shock wave against the washer will cause it to open slightly. The float is depressed into the water and subsequently bounces back to reclose the valve abruptly, creating a further shock wave which in turn rebounds to open the valve again; thus a cycle is established.

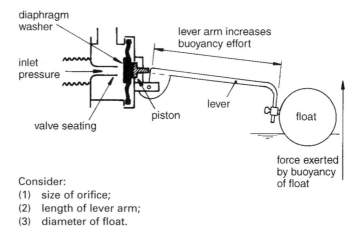

Consider:
(1)   size of orifice;
(2)   length of lever arm;
(3)   diameter of float.

As the cistern water level rises, the closing force exerted on the piston by the float and lever arm must close the diaphragm washer against the incoming force of the inlet pressure.

Figure 9.7   Prevention of float valve oscillation

Many plumbers suggest fitting a damping plate to the float or the lever arm, which will 'dampen down' the effects of the shock waves, or alternatively fitting baffle vanes in the cistern to prevent surface waves affecting the float. In practice these often seem a little 'Heath Robinson'.

Another way of improving flotation is simply to increase the lever arm length, thus achieving better leverage, or to fit a larger float. One could also reduce the size of the orifice seating. Although this might lead to more flow noise, it may be preferable to water hammer.

Tap washer oscillation

This is usually associated with worn, split or soft washers, which vibrate violently as water passes them (see figure 9.8). The noise can be extremely loud but the cure is simple – change the washer!

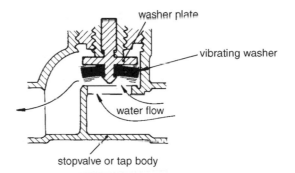

Worn, split or soft washers will vibrate violently.

Figure 9.8   **Tap and valve washer oscillation**

9.3   Other noises

Pumps

These will not cause excessive noise if well designed, unless flow exceeds the pump rating, or the static pressure is insufficient. Noise transmission from pumps can be reduced by:

- using rubber hose isolators between pump and pipework;
- isolating pipework from building structures with resilient inserts fitted in brackets;
- the use of a hydraulic–acoustic filter tuned to the unwanted frequency.

Splashing noises

These occur when water strikes the water surface in cisterns. Silencer tubes screwed into float valve outlets are now prohibited because of the risk of backsiphonage. Some float valves are fitted with collapsible silencer tubes, which are permitted. Others have outlets designed to reduce splashing noise.

Many taps are produced with flow correctors or aerators, which help to reduce water impact noise and splashing into sinks and other appliances. Metal sinks of pressed stainless steel increase splashing sounds, and some treatment in addition to that provided by the manufacturers may be worth considering.

Thermal movement

In hot pipes this causes creaks, squeaks and more impulsive sounds. The use of resilient pipe clips, brackets or pads between pipes and fittings will provide sufficient flexibility to cope with expansion and contraction. Long straight runs are a particular problem for which expansion joints or loops should be considered.

Air or vapour bubbles

These commonly cause flow noise. As a result of poor design and operation of hot water systems, bubbles are formed in hot water cylinders or heaters. Systems should be designed to avoid general or localized boiling and to allow removal of air when filling.

Gases

These are formed by corrosive action in primary circuits and can also cause noise when pumped around systems. This can normally be prevented in primary circuits by the use of corrosion inhibitors.

9.4   Noise transmission and reduction

Noise is transmitted to the listener from its source in a number of ways, including direct airborne transmission, through building structures (see figure 9.9), along pipes, and through the water contained in the pipe.

In metal pipes noise is transmitted with little loss, but plastics can reduce noise transmission depending on the pipe material and its thickness. For lengths of between 5 m and 20 m, the reduction is approximately 1 dB/m to 2.5 dB/m.

The insertion of metal-bellows vibration isolators can reduce transmissions by 5 dB to 15 dB, whilst reinforced rubber hose isolators can be even better.

Resilient mountings can help isolate a storage cistern from its supporting structure, preventing the transmission of noise through ceilings into habitable rooms.

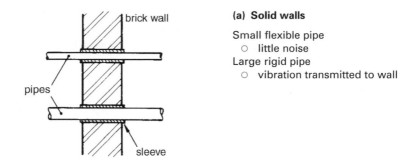

(a) Solid walls

Small flexible pipe
  ○ little noise
Large rigid pipe
  ○ vibration transmitted to wall

Brick will not transmit appreciable sound from small flexible pipes, but larger, more rigid pipes such as steel will induce some vibration.

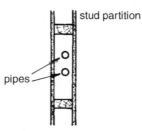

Hollow partition will act as a sound box to transmit and increase pipe noises.

Lightweight structures vibrate readily, transmitting and accentuating sound from pipes. To prevent this, use lightweight flexible pipes (copper or plastics) and flexible vibration isolating clips.

(b) Hollow walls

Figure 9.9   Structure-borne noises

Chapter 10
Accessibility of pipes and water fittings

For many years water byelaws stated that all pipes and fittings must be readily accessible for inspection, repair and renewal. Water Regulations now say:

'No water fitting shall be embedded in any solid wall or floor.'

While the regulations are, in essence, saying much the same as previous byelaws, the inference is that accessibility of fittings will be more strictly enforced.

However, the degree of accessibility will depend largely upon:

- how strictly the regulations are enforced by the water undertaker;
- the personal opinion of the designer, installer and building engineer;
- cost considerations of installation and maintenance of accessible installations using ducts, chases, access panels, etc.;
- the consequences of leakages from inaccessible parts of the pipework;
- the reliability of joints, resistance of pipes and joints to internal and external corrosion and flexibility of pipes when being passed through chased ducts or sleeves;
- the importance of aesthetic considerations, on the one hand where pipes are surface mounted and, on the other, the consequences of breaking into, repairing and otherwise spoiling expensive decorations and floor finishings.

10.1   Pipes passing through walls, floors and ceilings

No pipe may be installed in the cavity of an external cavity wall, other than where it has to pass through from one side to the other (see figures 10.1 and 10.2).

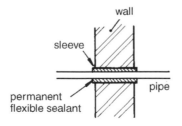

Pipes passing through walls or floors should be in a sleeve to permit movement relative to wall or floor.

Sleeves intended to carry a water pipe should not contain any other pipe or cable.

Figure 10.1   Pipe sleeved through wall

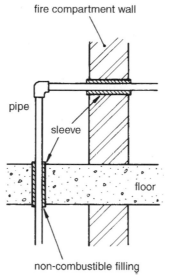

Sleeves should:

○ permit ready removal and replacement of pipes;
○ be strong enough to resist any external loading exerted by the wall or floor;
○ be sealed with fire-resistant material that will accommodate thermal movement.

Figure 10.2   Pipe sleeved through wall and floor

Pipes entering buildings

Pipes entering buildings should be located in a sleeve or duct that will permit the pipe to be readily replaced. Each end of the duct should be sealed using a non-hardening, non-cracking, water-resistant material for a length of 150 mm to prevent access by water, gas or vermin.

An example of pipes entering buildings is given in figure 10.3.

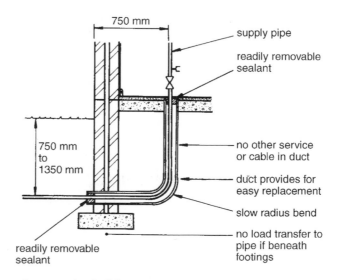

Figure 10.3   Access to pipes entering buildings

Pipes in walls, floors and ceilings

No pipe should be buried in any wall or floor, or under any floor, unless arranged as shown in figure 10.4. Where pipes are enclosed as shown it is preferred that there are no joints enclosed. Pipes should be wrapped to prevent corrosion and permit thermal movement.

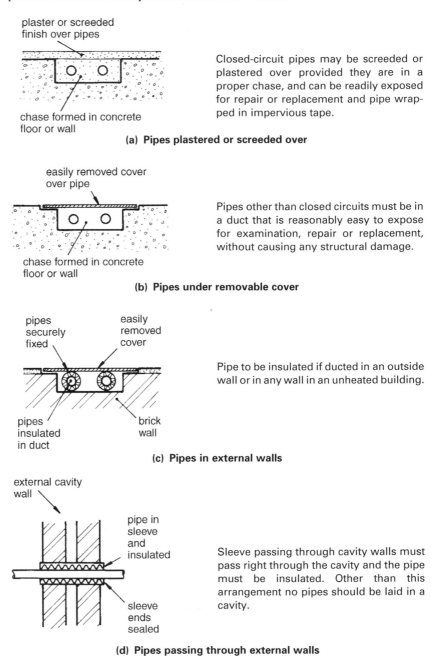

plaster or screeded
finish over pipes

chase formed in concrete
floor or wall

Closed-circuit pipes may be screeded or plastered over provided they are in a proper chase, and can be readily exposed for repair or replacement and pipe wrapped in impervious tape.

(a) Pipes plastered or screeded over

easily removed cover
over pipe

chase formed in concrete
floor or wall

Pipes other than closed circuits must be in a duct that is reasonably easy to expose for examination, repair or replacement, without causing any structural damage.

(b) Pipes under removable cover

pipes
securely
fixed

easily
removed
cover

pipes
insulated
in duct

brick
wall

Pipe to be insulated if ducted in an outside wall or in any wall in an unheated building.

(c) Pipes in external walls

external cavity
wall

pipe in
sleeve
and
insulated

sleeve
ends
sealed

Sleeve passing through cavity walls must pass right through the cavity and the pipe must be insulated. Other than this arrangement no pipes should be laid in a cavity.

(d) Pipes passing through external walls

Figure 10.4   Pipes in chases and ducts

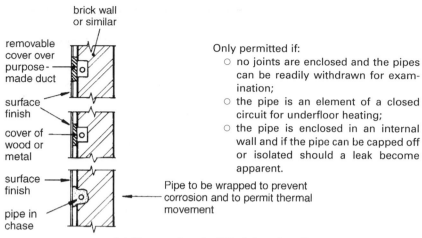

Only permitted if:
- ○ no joints are enclosed and the pipes can be readily withdrawn for examination;
- ○ the pipe is an element of a closed circuit for underfloor heating;
- ○ the pipe is enclosed in an internal wall and if the pipe can be capped off or isolated should a leak become apparent.

Pipe to be wrapped to prevent corrosion and to permit thermal movement

(e) Pipes enclosed within internal walls

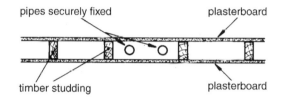

Pipes may be enclosed in internal walls of the timber studded type.

(f) Pipes in stud partition

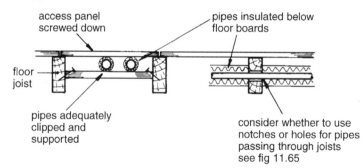

Pipes under suspended floors should be avoided. Where unavoidable they must be insulated and access panels formed in the floor for examination and repair.

Access panels should not affect the stability of the floor.

Boards removable at intervals of not more than 2 m and at every joint for inspection of whole length of pipe.

(g) Pipes under timber floors

Figure 10.4
continued
Pipes in chases and ducts

continued

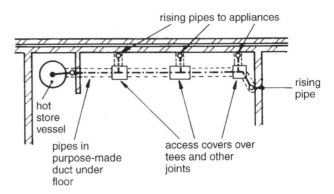

Where pipes are enclosed in a solid floor, access covers should be provided at tees and joints.

(h) Typical ducting arrangements

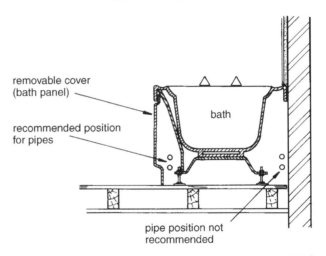

Although BS 6700 recommends that pipes are positioned in front of a bath, the back position gives better fixing for pipes and easier access for bath replacement. However, joints behind bath should be avoided.

(i) Pipes concealed under baths

Figure 10.4 **Pipes in chases and ducts**
continued

10.2 Stopvalves

Stopvalves above ground should be positioned so as to be readily accessible for examination, maintenance and operation.

Stopvalves on underground pipes should be enclosed within a pipe guard or chamber with surface box to provide access for shutting off with a metal stopvalve key (see figure 10.5).

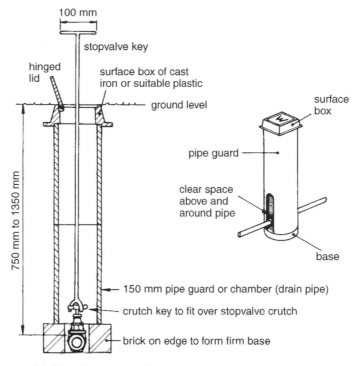

(a) Chamber construction

Surface box to be suitable for relevant traffic loading, e.g. heavy or light grade.

Stopvalves and pipes more than 1350 mm deep will be deemed to be inaccessible.

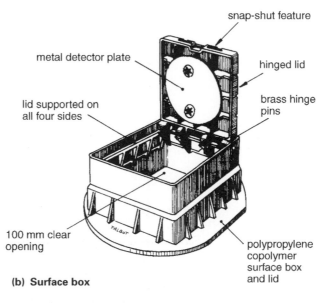

(b) Surface box

Figure 10.5   Access to below ground stopvalves

10.3 Water storage cisterns

These should be positioned so that they are easy to clean and maintain (see figure 10.6).

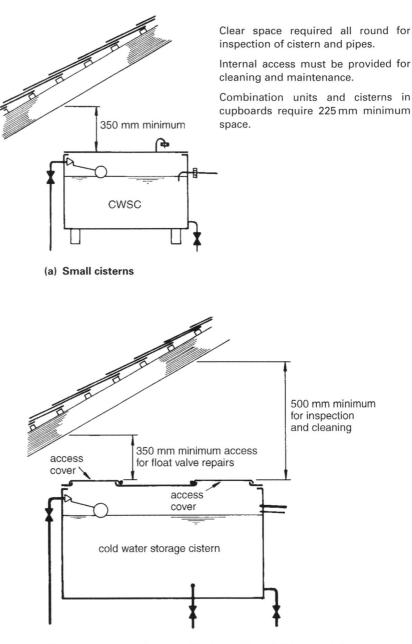

Clear space required all round for inspection of cistern and pipes.

Internal access must be provided for cleaning and maintenance.

Combination units and cisterns in cupboards require 225 mm minimum space.

350 mm minimum

CWSC

(a) Small cisterns

500 mm minimum for inspection and cleaning

350 mm minimum access for float valve repairs

access cover

access cover

cold water storage cistern

Clear space required all round for inspection of cistern and pipes.

(b) Large cisterns

Figure 10.6   Access to cisterns

Chapter 11
Installation of pipework

Pipes and fittings need to be chosen and installed to suit their purpose and the conditions in which they are to be situated.

Pipes should be laid or fixed:

- to avoid frost, mechanical damage or corrosion;
- so that they do not leak, cause undue noise or permit any contamination of water contained in them;
- to comply with the requirements of relevant British Standards or be approved under the UK Water Fittings Testing Scheme.

When jointing pipes take care to ensure that:

- joints are mechanically sound and clean inside;
- pipes are cut squarely and all burrs removed and distorted ends rounded;
- cutting tools are in good condition to limit distortion;
- joints comply with British or European Standards and are listed in the *Water Fittings and Materials Directory*;
- when applying heat the risk of fire is eliminated, and the operator does not breathe any harmful fumes given off from soldering or welding processes;
- only approved jointing materials or compounds are used;
- joints, clips and fittings are compatible with the pipe material and will not cause corrosion.

Pipes and components should be handled with care. Damage caused during installation can seriously affect the life and performance of the system.

Bending of pipes should be carefully carried out using purpose-made equipment to avoid deformation or damage. Avoid crimping, kinking and restriction of the pipe bore which can cause damage to the pipe bore, or loss of water flow, or create additional flow noise in the system. Any damaged pipes should be discarded. Pipes of copper, stainless steel and black low carbon steel are ideally suited to bending. Pipes of galvanized steel must *not* be bent as this will damage the protective zinc coating.

11.1 Steel pipes

There are three grades of low carbon steel pipes to BS 1387: heavy, medium and light, each of which is obtainable with or without galvanization inside and out. For hot and cold services, only galvanized pipes are permitted, as follows:

Heavy – (identified by a red band painted around the pipe near its ends) for use underground where, in addition to galvanization, other forms of protection should be used to guard against exterior corrosion, e.g. a bituminous coating. Exposed threads should be painted.

Medium – (identified by a blue band) for general use above ground only.

Light – (identified by a brown band) permitted on some fire fighting installations.

Jointing of galvanized steel tubing for water services is usually achieved by screwing the pipe and fitting together, pipe joints being made on site using hand or electrically powered threading machines. After jointing, any exposed threads should be painted, of if underground, treated with a suitable bitumen or corrosion-preventing coating.

Low carbon steel pipes and fittings should be protected from corrosion, particularly in damp or otherwise corrosive conditions such as below ground. Proprietary pipe wrap materials are available for this purpose, and should be applied so that none of the pipework remains exposed.

It is advisable to consult water suppliers before installing any low carbon steel to establish whether the water supply will cause excessive corrosion.

A selection of joints for use with low carbon steel tube can be seen in figures 11.1–11.3, whilst further information on jointing and fixing is given in tables 11.1 and 11.2.

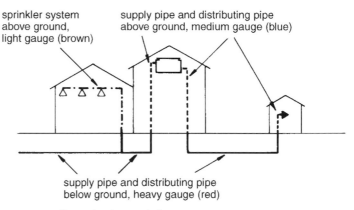

sprinkler system above ground, light gauge (brown)

supply pipe and distributing pipe above ground, medium gauge (blue)

supply pipe and distributing pipe below ground, heavy gauge (red)

Control valves not shown.

Figure 11.1   Grades of steel pipe and their uses

Table 11.1   **Thread engagement lengths for steel pipe to BS 1387**

Nominal size of pipe inside diameter mm	Thread length mm
15	13
20	15
25	17
32	19
40	19
50	24
80	30
100	36
150	40

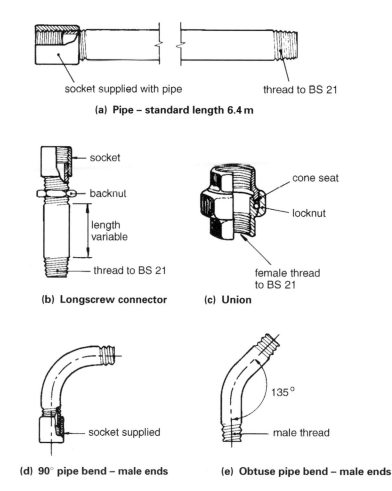

socket supplied with pipe   thread to BS 21

(a) Pipe – standard length 6.4 m

socket

backnut

length variable

thread to BS 21

(b) Longscrew connector

cone seat

locknut

female thread to BS 21

(c) Union

socket supplied

(d) 90° pipe bend – male ends

135°

male thread

(e) Obtuse pipe bend – male ends

Figure 11.2   **Selection of fittings for steel tubes to BS 1387**

continued

(f) 90° elbow female ends

female ends

female
end

male end

(g) Obtuse bend – male and female

(h) Tee 90° – equal

Numbers indicate method of specifying outlets:
- ○ state 'in line' outlets first;
- ○ state larger end first.

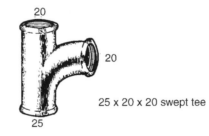

20

20

25 x 20 x 20 swept tee

25

(i) Swept or pitched tee

Figure 11.2 continued   **Selection of fittings for steel tubes to BS 1387**

Table 11.2   **Maximum spacing of fixings for internal steel piping**

Nominal size of pipe inside diameter mm	Spacing on horizontal run m	Spacing on vertical run m
15	1.8	2.4
20	2.4	3.0
25	2.4	3.0
32	2.7	3.0
40	3.0	3.6
50	3.0	3.6
80	3.6	4.5
100	3.9	4.5
150	4.5	5.4

(a) Saddle clip of galvanized steel

(b) Screw-on clip of galvanized cast iron

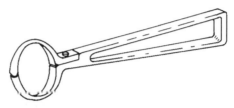

(c) Clip of galvanized cast iron for building in

Figure 11.3   Pipe fixings for steel tubes

11.2   Copper pipes

Copper tube is ideally suited to use in hot and cold water supplies. It is resistant to external corrosion in most soil conditions and to internal corrosion from the majority of water supplies. Copper is high on the electrochemical scale and should not be connected directly to other metals, particularly galvanized steel, unless the other metal is resistant to, or protected from, the effects of galvanic (electrolytic) action. In areas where unacceptable green staining occurs, or electrolytic corrosion is promoted, alternative materials should be chosen, or water treatment should be considered.

Only copper tube to BS EN 1057 should be used. This standard includes those grades of tube that were formerly covered by BS 2871.

Copper tube for water services to BS EN 1057 is graded according to its hardness in the range of sizes shown in Table 11.3.

Table 11.3   **Copper tube to BS EN 1057**

Material temper		Range of sizes (OD)		Standard lengths available
EN hardness number	Common term	mm from	mm to	m
R220	Annealed	6	54	10 to 20
R250	Half hard	6	159	3 and 6
R290	Hard	6	267	3 and 6

OD = outside diameter.

Annealed (soft temper) tube is used from coils for microbore heating circuits, smaller connections to taps and in a heavier wall thickness for underground use.

Half hard tube is obtainable in straight lengths in thicknesses suitable for above and below ground use. It is ideally suited to bending by machine or by hand.

Hard tube is obtainable in straight lengths. Because of its hardness it will not readily stretch and should not be bent.

Tables 11.4 and 11.5 give a range of pipes to BS EN 1057 and show how the former BS 2871 tube has been absorbed into this new European Standard. Typical fixings for copper tube are illustrated in figures 11.4–11.6 and spacing recommendations are given in table 11.6.

Table 11.4   **R250 half hard copper tube to BS EN 1057 in straight lengths**

OD mm	Wall thickness								
	0.6	0.7	0.8	0.9	1	1.2	1.5	2	2.5
6	X		Y						
8	X		Y						
10	X		Y						
12	X		Y						
15		X			Y				
22				X		Y			
28				X		Y			
35						X	Y		
42						X	Y		
54						X		Y	
66.7						X		Y	
76.1							X	Y	
108							X		Y
133							X		
159								X	

Notes
(1)   X indicates tube to former BS 2871: Part 1, Table X.
(2)   Y indicates tube to former BS 2871: Part 1, Table Y.

OD = outside diameter.

Table 11.5   **R220 annealed copper tube to BS EN 1057 in coiled lengths**

OD mm	Wall thickness					
	0.6	0.7	0.8	0.9	1	1.2
6	W		Y			
8	W		Y			
10		W	Y			
12			Y			
15					Y	
22						Y
28						Y

Notes
(1)   W indicates tube to former BS 2871: Part 1, Table W.
(2)   Y indicates tube to former BS 2871: Part 1, Table Y.

OD = outside diameter.

Copper tube is also available with a polyethylene coating to add protection from external corrosion. Joints would need wrapping with a suitable adhesive tape after testing. In some areas this may be necessary to prevent corrosion to underground pipes. The polyethylene coating should be coloured blue when the tube is to be used underground.

Table 11.6   **Maximum spacing of fixings for internal copper and stainless steel piping**

Typing of piping	Nominal size of pipe, outside diameter mm	Spacing on horizontal run m	Spacing on vertical run m
Copper tube to	15	1.2 (1.8)*	1.8 (2.0)*
BS EN 1057	22	1.8 (2.4)	2.4 (3.0)
R250 half hard,	28	1.8 (2.4)	2.4 (3.0)
R290 hard	35	2.4 (2.7)	3.0 (3.0)
and	42	2.4 (3.0)	3.0 (3.6)
stainless steel to	54	2.7 (3.0)	3.0 (3.6)
BS 4127: Part 2	76	3.0 (3.0)	3.6 (3.6)
	108	3.0 (3.6)	3.6 (4.5)
	133	3.0 (3.9)	3.6 (4.5)
	159	3.6 —	4.2 —

Note   * Figures for stainless steel tube are shown in brackets.

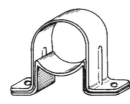
(a) Two-piece spacing clip

(b) Single-piece spacing clip

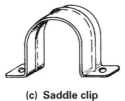
(c) Saddle clip

Figure 11.4   **Copper fixing clips for small diameter copper tubes**

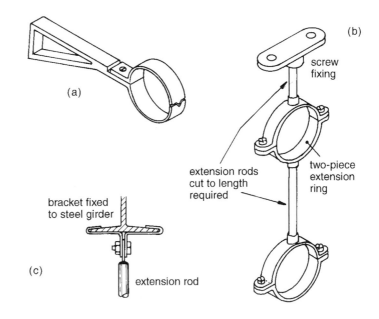

(a) **Built-in clip**

(b) **Two-piece pipe ring with extension rod and backplate**

(c) **Alternative fixing for pipe ring**

Figure 11.5   **Brass fixing clips for copper tubes**

(a) Snap-fit spacer clip in PVC-U

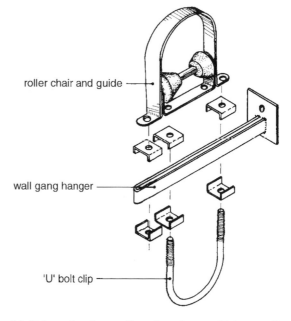

roller chair and guide

wall gang hanger

'U' bolt clip

(b) Fixings for large diameter pipes which permit expansion to take place

Figure 11.6   Other fixings for copper tubes

Jointing methods for copper tubes

Copper fittings may be manufactured from copper or copper alloys such as brass or gunmetal. Copper is used for capillary solder fittings only. Fittings made from copper alloys should comply with the requirements of BS EN 1254.

For underground use, and in situations where the water is capable of causing dezincification, fittings are required to be dezincification-resistant. Dezincification is a form of electrolytic corrosion in which the zinc content of the brass is corroded away, leaving the copper behind and the remaining brass in a porous and weakened condition.

Gunmetal, made of copper and tin, contains no zinc and is, therefore, immune to dezincification. It is, however, expensive to manufacture because each individual fitting must be individually cast. Brass on the other hand can be hot stamped into shape before machining.

Duplex brass has traditionally been used for the manufacture of fittings, but it was susceptible to dezincification and its use was permitted only above ground in those areas not likely to be affected by dezincification. Duplex brass has now been replaced by dezincification-resistant (DZR) brass for the production of the vast majority of copper alloy fittings. DZR brass fittings are marked with the dezincification symbol:

Suitable methods for the jointing of copper tube include: compression fittings, capillary solder joints, brazing and braze welding. These are described below.

Compression fittings – type A (non-manipulative)

These are the most usual type of compression fitting for use above ground. They must not be used below ground. See figures 11.7 and 11.8.

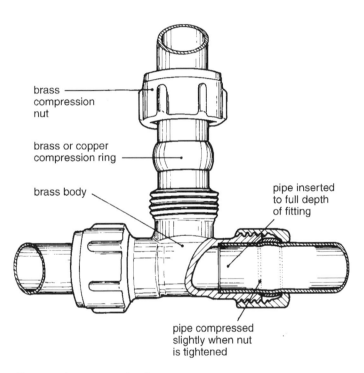

brass compression nut

brass or copper compression ring

brass body

pipe inserted to full depth of fitting

pipe compressed slightly when nut is tightened

For use above ground only.

Also suitable for stainless steel tube.

Cut pipe ends squarely and file off burrs.

No manipulation of pipe ends needed.

Do not overtighten brass backnut.

Jointing compounds not required.

(a) Assembly of compression fittings – type A (non--manipulative)

Figure 11.7   Compression fittings – type A

continued

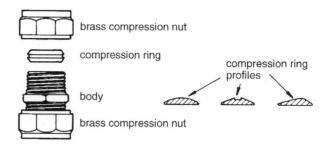

brass compression nut

compression ring

compression ring profiles

body

brass compression nut

Rings from one fitting are not compatible with fittings from another manufacturer.

(b) Straight coupling

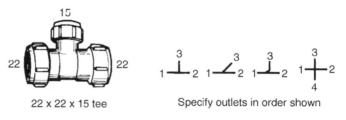

22 x 22 x 15 tee

Specify outlets in order shown

(c) Tee with reduced branch

(d) Elbow – copper to male iron

(e) Straight swivel coupling
Supplied with one fibre washer

Figure 11.7
continued

Compression fittings – type A

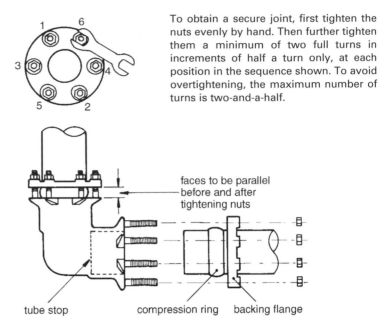

To obtain a secure joint, first tighten the nuts evenly by hand. Then further tighten them a minimum of two full turns in increments of half a turn only, at each position in the sequence shown. To avoid overtightening, the maximum number of turns is two-and-a-half.

faces to be parallel before and after tightening nuts

tube stop   compression ring   backing flange

Figure 11.8   **Large diameter compression fitting – type A**

Compression fittings – type B (manipulative)

These require the flaring of pipe ends and give extra security against joints pulling apart in extreme conditions. This is especially important underground, where any resulting leakage could well go undetected. See also figures 11.9 and 11.10.

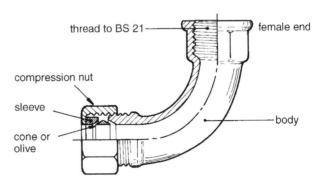

thread to BS 21 — female end

compression nut

sleeve

cone or olive

body

Figure 11.9   **Type B compression fitting for copper tube – female iron to copper bend**

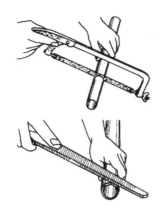

(1) Cut pipe squarely.

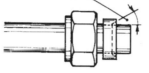

(2) File ends to remove burrs.

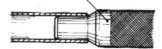

cone angle

(3) Slip backnut and sleeve on to end of pipe.

flaring tool or drift

(4) Flare out tube end.

The flaring tool from one manufacturer should not be used with another make of fitting because the flare angle may not match the olive and chamfered fitting.

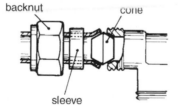

backnut   cone

sleeve

(5) Fit pipe to fitting with cone inserted.

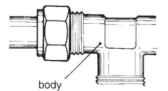

body

(6) Screw up backnut hand tight, then tighten a further one-and-a-half turns.

Figure 11.10   **Method of assembly of type B compression fittings**

Capillary fittings

These are made in two types, as illustrated in figure 11.11. They are suitable for use both above and below ground and are made in a range of copper or brass materials.

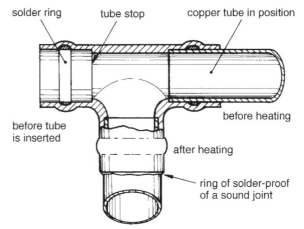

Extra solder should not be added as there is already sufficient within the fitting

(a) Integral solder ring type

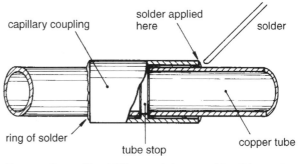

Because the end feed fitting contains no solder, this must be applied to the joint as it is heated.

(b) End feed type

Figure 11.11   Capillary soldered joints

The soldering method is as follows:

(1) Cut the end square and remove burrs.
(2) Thoroughly clean inside of fitting and outside of pipe with steel wool or sandpaper (not emery board).
(3) Apply suitable flux to prevent oxidation and assist cleaning (just a thin smear to both surfaces).
(4) Assemble joint.
(5) Apply heat with blow torch or electrical tongs.
(6) Add solder to joint area (end feed type only) and continue to apply heat until solder flows around and into joint. For solder ring fittings apply heat until solder appears and forms a ring around the end of the fitting (no extra solder needed).
(7) Leave joint undisturbed to cool.
(8) Remove flux residues or they may corrode the pipe.

Solders have traditionally been lead/tin alloys, the use of which is prohibited by Water Regulations in favour of lead-free solders, e.g. tin/silver alloy.

Brazed joints

Brazing is a capillary joint which uses a filler metal that has a melting point above 450°C but which is lower than the melting point of the metals being joined. The method of jointing is similar to that of soft soldering but it is carried out at higher temperatures. In brazing, the parent metals are not melted, rather they are 'wetted' or 'tinned' by elements within the filler metal.

Brazing is ideally suited to joints that need to resist high pressures or temperatures, and in the past has been used extensively in hospitals where specifications have demanded the use of capillary joints without a lead content. Joints may be hand-made using specialized equipment or by the use of copper or copper alloy fittings, similar in appearance to capillary soldered joints. Further information on brazed joints is given in figure 11.12 and tables 11.7 and 11.8.

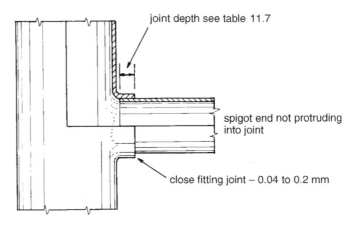

Figure 11.12   **Hand-made brazed joint to copper tube**

Table 11.7   **Minimum depths of overlap for brazed joints**

Pipe diameter (mm)	15	22	28	35	42	54	67	76	108
Joint depth (mm)	8	10	12	15	18	20	22	25	30

Brazing filler metals may be of two types:

(1)   Copper-silver-zinc alloys (Cu-Ag-Zn)
(2)   Copper-phosphorus-silver alloys (Cu-P-Ag)

These are shown in table 11.8 which is based on BS 1845.

Table 11.8   Brazing alloys

(a) Group AG: Silver brazing filler metals

BS type	Nominal percentage composition (%)						Melting range °C	
	Silver	Copper	Zinc	Cadmium	Tin	Nickel	Solidus	Liquidus
AG1	50	15	16	19	0	0	620	640
AG2	42	17	16	25	0	0	610	620
AG7	72	28	0	0	0	0	670	780
AG9	50	15.5	15.5	16	0	3	635	655
AG13	60	26	26	0	0	0	695	730
AG14	55	21	21	0	2	0	630	660
AG15	44	30	30	0	0	0	675	735

Notes
(1) Cadmium can produce highly toxic fumes and the use of alloys containing cadmium is not recommended.
(2) Zinc is also toxic and alloys containing zinc should be used with care, with the work area well ventilated.

(b) Group CP: Copper-phosphorus brazing filler metals

BS type	Nominal percentage composition (%)			Melting range 8C	
	Copper	Phosphorus	Silver	Solidus	Liquidus
CP1	Remainder	4.6	14.5	645	700
CP2	Remainder	6.5	2	645	740
CP3	Remainder	7.6	0	705	800
CP4	Remainder	6	5	640	740

Notes
(1) CP alloys may be used on copper to copper joints without the need for a flux.
(2) CP alloys should never be used on ferrous metals or alloys containing nickel.

Fluxes for brazing may be of two types:

(1)   for high temperature brazing (over 750°C) a borax based flux should be used;
(2)   for low temperature brazing (below 750°C) an alkali fluoride based flux should be used.

Braze-welded joints

Braze welding is described in BS 699 as 'the jointing of metals using a technique similar to fusion welding and a filler metal with a lower melting point than the parent metal, but neither using capillary action as in brazing nor intentionally melting the parent metal'.

Braze (bronze) welded joints (see figure 11.13) are very strong, but are costly to produce and require the use of a high degree of skill.

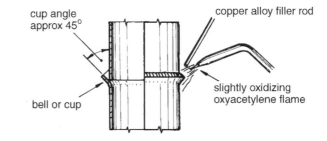

cup angle approx 45°	copper alloy filler rod
bell or cup	slightly oxidizing oxyacetylene flame

Figure 11.13   Braze-welded butt joint

Fluxes used should be borax based and should be cleaned off after use. Filler metals will be of brass with the addition of silicon and tin. Two suitable metals are listed in BS 1453 and BS 1845. See table 11.9.

Table 11.9   Recommended filler metals for braze-welding copper joints

Type		Nominal percentage composition (by weight)					
BS 1453	BS 1845	Copper	Zinc	Silicon	Tin	Magnesium	Iron
C2	CZ26	57 to 63	Balance	0.02 to 0.5	Optional to 0.5	0	0
C4	CZ27	57 to 63	Balance	0.15 to 0.3	Optional to 0.5	0.05 to 0.25	0.1 to 0.5

11.3   Stainless steel pipes

Stainless steel pipe to BS 4127 is similar to copper in use. Its external dimensions are the same as copper. The fittings used are generally those made for copper tubes to BS 864: Part 2.

However, stainless steel pipe is much more rigid than copper and therefore needs a little more accuracy in bending. Bending is done using the copper pipe bending machine.

Pipe diameters range from 15 mm to 159 mm. Spacings for fixings are given in table 11.6.

A selection of joints for stainless steel pipes is shown in figure 11.14.

Although stainless steel and copper pipes may generally be mixed, small copper areas to large stainless steel areas should be avoided.

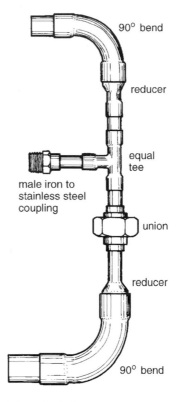

Figure 11.14   **Selection of joints for stainless steel pipes**

Jointing methods are similar to those used for copper tubes:

(1)   compression fittings of copper alloy or stainless steel;
(2)   capillary soldered joints of copper alloy or stainless steel. For soldered joints use phosphoric acid based flux;
(3)   silver soldering and brazing;
(4)   anaerobic adhesive bonding (up to 85°C) subject to the following restrictions in Note (1) (see figure 11.15).

Note (1)   No metal pipes shall be connected by means of adhesive jointing where the pipe may be:
(a)   embedded in a wall or floor:
(b)   enclosed in a chase or duct; or
(c)   in a position where access is restricted.

Note (2)   For methods (1) to (3) see jointing methods for copper tube. See also figure 11.10.

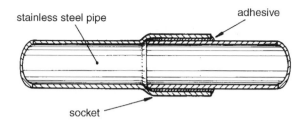

Method for anaerobic adhesive bonding
(1) Check fit between tube and fitting.
(2) Ensure bond areas are grease-free using solvent degreaser.
(3) Abrade bonding surfaces with medium emery cloth (80 grit).
(4) Apply ring of adhesive to leading edge of pipe and slip pipe into fitting.
(5) Allow:
    ○ one minute for curing,
    ○ one hour to withstand static pressure,
    ○ one day for full strength.

Figure 11.15   **Anaerobic adhesive bonded joint for stainless steel tubes**

11.4   Plastics pipes

There are a number of plastics pipes in use for hot and cold water supply installations. These are considered separately in the following pages.

It is important that the various plastics are recognized individually as separate materials, e.g. polyethylene (PE), unplasticized polyvinyl chloride (PVC-U), chlorinated polyvinyl chloride (PVC-C), cross-linked polyethylene (PE-X), etc. Each material has its own range of properties and jointing methods, and is often suited to different applications and uses.

There is one area, however, where all the plastics pipe materials considered here can, to some degree, be considered together, and that is their reaction to the effect of heat. Coefficients of expansion in plastics are generally about ten times greater than for metals, so greater attention should be given to the allowances made for thermal movement.

Plastics pipes have relatively low softening and melting points, much lower than the metals. They should, therefore, not be used where heat may cause them to become softened and weakened.

Plastics pipes used for hot water must be capable of withstanding temperatures of 100°C or more.

Plastics pipes are flexible and may tend to droop between fixings, particularly where clips are too widely spaced.

It is generally recommended that plastics pipes are not threaded because it reduces the pressure resistance of the pipe. If threaded joints are desired, the manufacturer should first be consulted as to the advisability of their use.

Where jointing materials or packing are required, a PTFE (polytetrafluoroethylene) tape is recommended. Oil based compounds should not be used as they may deteriorate the plastics material of the pipe or its fittings. Steel grips or stilsens that may damage the pipe should not be used on plastics pipes and fittings.

Plastics pipework systems are not automatically intercompatible, and it is recommended that differing plastics are not mixed within a pipework system.

Polyethylene pipes (PE)

Polyethylene has proved to be an excellent material for cold water installations, particularly when used below ground. It is corrosion-resistant, easy to lay and simple to joint. It can be obtained in long lengths permitting supply pipes to be laid with the minimum of joints. Its flexibility allows it to be bent around obstacles and threaded through ducts into buildings. It must not be used for hot water installations.

When laying polyethylene pipe below ground, it is advisable to bed and cover the pipe with a selected soil, or granular material such as pea shingle, to prevent damage from stones and flints, and to avoid deformation of the pipe when backfilling the trench.

In the past polyethylene has been known to be permeable to gas, and is liable to be damaged by contact with oils or oil based products.

For use in construction work, polyethylene tube is manufactured in two types:

- Blue medium-density polyethylene tube to BS 6572 is made for use below ground, or in positions where the tube is fully protected from sunlight, e.g. within ducts. Its blue colour makes it easy to identify below ground.
- Black medium-density polyethylene tube to BS 6730 is made for above ground use. The tube material has a carbon black pigment added to prevent the passage of light through the pipe wall.

Additionally there are a number of other polyethylene pipes in use meeting various previous standards. Table 11.10 gives information on some of these. Care should be taken when connecting these to existing pipelines that any fittings and inserts used are correct for the dimensions of the particular pipe.

Table 11.10   Types and grades of polyethylene pipes

Type of piping	Grade	Maximum working pressure	Type of piping	Nominal size, outside diameter mm	Approximate bore mm	Maximum working pressure bar
Low-density polyethylene (Type 32) to BS 1972 and high-density polyethylene (Type 50) to BS 3284	Class B (light)	6	Blue medium-density polyethylene to BS 6572 for below ground use only and black medium-density polyethylene to BS 6730 for above ground use only	20	15	
				25	20	
	Class C (medium)	9		32	25	12
				50	40	
	Class D (heavy)	12		63	50	

Preferred lengths: straight pipes 6 m, 9 m, 12 m.   Coils 50 m, 100 m, 150 m.

Note   BS 1972 is obsolescent and BS 3284 is withdrawn. These are shown because of their extensive use in the past.

Fixing distances for polyethylene pipes

Because polyethylene is a flexible material, pipes above ground should be supported continuously, or at least be fixed to the spacing requirements shown in table 11.11.

Table 11.11   Maximum spacing of fixings for internal polyethylene pipes

Type of piping	Nominal size of pipe	Spacing on horizontal run m	Spacing on vertical run m
Low-density polyethylene to BS 1972 (nominal size in inches)	$\frac{3}{8}$	0.30	0.60
	$\frac{1}{2}$	0.40	0.80
	$\frac{3}{4}$	0.40	0.80
	1	0.40	0.80
	$1\frac{1}{4}$	0.45	0.90
	$1\frac{1}{2}$	0.45	0.90
	2	0.55	1.10
	$2\frac{1}{2}$	0.55	1.10
	3	0.60	1.20
	4	0.70	1.40
Medium-density polyethylene to BS 6730 (black) (nominal size in millimetres)	20	0.5	0.9
	25	0.6	1.2
	32	0.6	1.2
	50	0.8	1.5
	63	0.8	1.6

Jointing methods for polyethylene pipes

A variety of jointing methods are available for polyethylene pipes and these are produced from both plastics and metals. For smaller pipes, mechanical joints are predominantly used, whereas for larger sizes, and particularly for mains pipes, thermal fusion jointing methods are currently popular.

BS 6700 recommends that the requirements of CP 312: Part 3 be followed when installing and jointing polyethylene pipes. This standard was, however, written in 1973 and is likely to be replaced in the future by European standards.

Compression fittings

These are shown in figures 11.16–11.18.

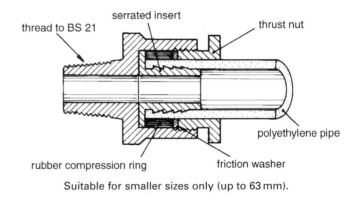

Suitable for smaller sizes only (up to 63 mm).

Figure 11.16   Compression fitting for polyethylene pipes

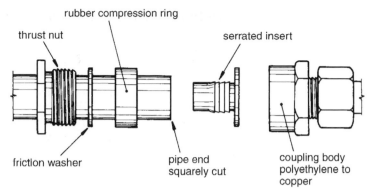

This illustration and the instructions for making the joint are for a compression joint manufactured by Talbot. Similar fittings are available from several other manufacturers.

(1)   Asemble parts in sequence shown.
(2)   Using a hide mallet or similar, knock insert into pipe end until its flange touches pipe face.
(3)   Push rubber compression ring and friction washer against flange and locate assembly in fitting body.
(4)   Screw in thrust nut until hand tight, then tighten fully with spanner (one-and-a-half to two turns).

Figure 11.17   Making the compression joint

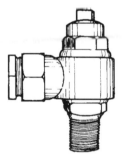

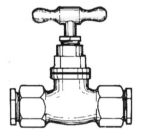

(a) Under pressure tapping ferrule   **(b) Stopvalve**

Figure 11.18   Examples of fittings for polyethylene pipes

Push-fit joints

These are shown in figures 11.19–11.22.

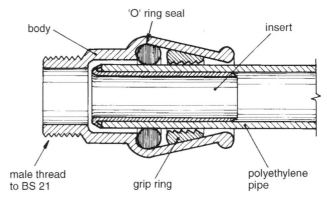

Figure 11.19   **Push-fit joint for polyethylene pipe**

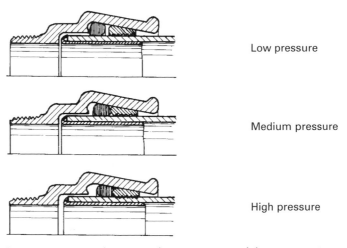

As water pressure increases the components tighten to create a pressure seal.

Figure 11.20   **How the push-fit joint works**

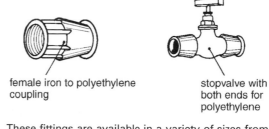

female iron to polyethylene
coupling

stopvalve with
both ends for
polyethylene

These fittings are available in a variety of sizes from
20 mm to 180 mm.

Figure 11.21   **Examples of small diameter fittings**

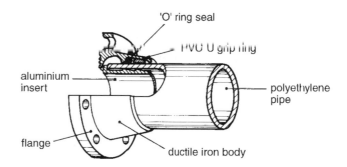

'O' ring seal

PVC U grip ring

aluminium
insert

polyethylene
pipe

flange

ductile iron body

Figure 11.22   **Example of large diameter fitting – flange adaptor**

Thermal fusion joints

There are three main techniques currently in use for the making of thermal fusion joints: socket fusion, butt fusion and electrofusion. These are shown in figures 11.23–11.25. It is important that heat input is strictly controlled to give correct fusion temperatures. Operatives should be fully trained in thermal fusion techniques.

Thermal fusion joints are suitable for large diameter polyethylene (PE) pipes, but are not popular for small diameters because other methods of jointing are so much easier. They are also suitable for polypropylene (PP) pipes.

(a) Socket fusion

Pipes may be joined using a range of polyethylene fittings. Pipe ends are heated to soften mating surfaces using a heating element, after which the pipe end is pushed fully into the socket and held firmly until fused and cooled down.

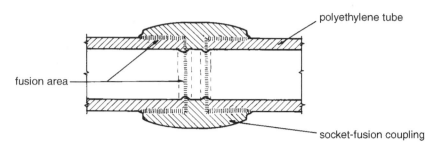

(b) Butt fusion

Used to join pipe ends only. Jointing is achieved by softening ends using a flat plate heating element, following which the pipe ends are pressed firmly together to form one solid mass. Popular method for the jointing of water mains.

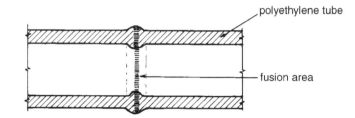

(c) Electrofusion

Perhaps the most reliable method but more expensive. A range of fittings are available, e.g. tees, bends, couplings, tapping saddles, etc. Fittings have an electric heating coil built in to each socket. The pipe is inserted into the socket, an electrical connection is made and the heat automatically applied. The joint is allowed to cool after the correct fusion temperature has been reached.

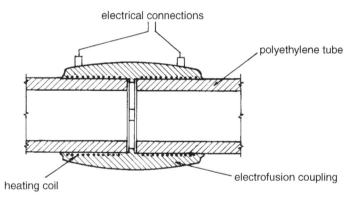

Figure 11.23   Thermal fusion jointing methods

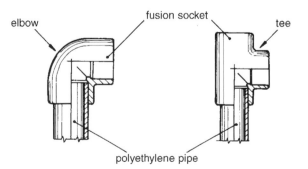

Figure 11.24   Thermal fusion joints

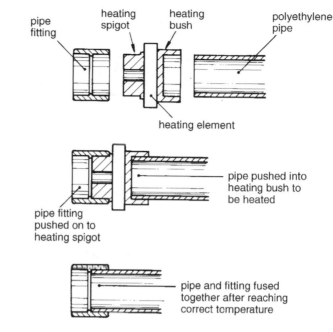

Also suitable for polypropylene pipe.

Pipes should only be joined with compatible fittings, i.e. pipe and fittings in polyethylene or pipe and fittings in polypropylene.

Correct heat is essential.

Follow manufacturers' instructions for jointing method.

Figure 11.25   **Making socket fusion joints**

Unplasticized polyvinyl chloride (PVC-U) pipes

PVC-U (unplasticized polyvinyl chloride to BS 3505) is an excellent material for cold water pipes for temperatures up to 20°C, beyond which its mechanical properties are reduced. Available in three pressure ranges and a variety of diameters, from 10 mm to 600 mm (see table 11.12), it is used in great quantities for water mains but is not so popular for smaller services. It is equally suited to all installations above and below ground, for cold water only. A similar material, PVC-C (chlorinated polyvinyl chloride), is suitable for use with hot water.

Table 11.12   **Sizes and pressure ratings of PVC-U pipe to BS 3505**

Pressure class	Pressure rating at 20°C bar	Range of nominal sizes, inside diameter inches	mm
C	9	2 to 24	50 to 600
D	12	$1\frac{1}{4}$ to 18	32 to 450
E	15	$\frac{3}{8}$ to 16	10 to 400

Fixings for PVC-U pipes

Examples of fixings and their spacing are shown in figure 11.26 and table 11.13.

(a) **Snap-fit clip**

(b) **Saddle clip**

(c) **Galvanized steel valve support plate (to fit valve flange)**

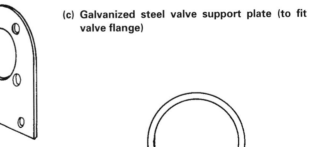

(d) **Polyethylene clip with metal building-in bracket**

Figure 11.26   **Fixings for PVC-U pipes**

Table 11.13   Maximum spacing of fixings for internal PVC-U and PVC-C pipes

Type of piping	Nominal size of pipe inches	mm	Spacing on horizontal run m	Spacing on vertical run m
PVC-U to BS 3505 (Figures are for normal ambient temperatures below 20°C. For temperatures above 20°C the pipe manufacturer should be consulted)	$\frac{1}{4}$	10	0.6	1.1
	$\frac{1}{2}$	13	0.7	1.3
	$\frac{3}{4}$	20	0.7	1.4
	1	32	0.8	1.6
	$1\frac{1}{4}$	32	0.9	1.7
	$1\frac{1}{2}$	40	1.0	1.9
	2	50	1.1	2.2
	3	75	1.4	2.8
	4	100	1.6	3.1
	6	150	1.9	3.7
PVC-C (Based on average temperature of 80°C)		12 to 25	0.5	1.0
		32 to 63	0.8	2.2

Jointing methods for PVC-U pipes

For below ground use it is recommended that mechanical joints are used in preference to solvent welding, particularly where conditions are wet and muddy. Any metal fittings should be immune to dezincification.

Threaded joints

These should not be used for PVC U pipes to BS 3505. However, this method can be used for pipe to BS 3506 class 7 up to 2 in diameter, providing the pressure does not exceed 9 bar.

Compression joints

See figure 11.27. These are similar to those for polyethylene. BS 6700 suggests that joints must be made of PVC-U also, but there is at least one manufacturer who produces approved fittings of brass and gunmetal. These are suitable for joining tubes of up to 50 mm diameter.

 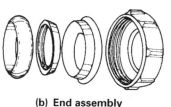

(a) **Coupling body**   (b) **End assembly**

Similar fitting can be used for polyethylene pipes

Care should be taken not to overtighten joints.

Figure 11.27   Compression joint for PVC-U pipes

Mechanical joints

See figure 11.28 as an example.

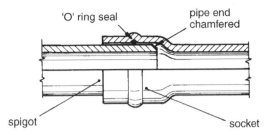

The push-fit joint for PVC-U is the simplest and cheapest form of installation.

Mechanical joints of push-fit type to BS 4346: Part 2 are suitable for use on pipes of 50 mm diameter and above.

Care should be taken to ensure that bends and branches are secure and properly anchored before pressure is applied.

Pipe end should be lubricated with suitable bactericidal, non-toxic lubricant to assist insertion into socket.

Pipe end should be inserted fully into the socket, making any allowance for expansion as recommended by the manufacturer.

Figure 11.28 **Mechanical joint for PVC-U pipe (push fit type)**

Solvent cement joints

Solvent-welded joints are commonly used for the jointing of PVC-U pipes. For below ground use it is recommended that mechanical joints are used in preference to PVC-U, particularly where conditions are wet and muddy. Where solvent welded joints are used on long pipelines, thermal movement should be accommodated by the use of the occasional mechanical or push-fit joint. Failures with solvent cement jointing will nearly always result from operatives not adhering to manufacturers' jointing procedures.

Solvents used should comply with BS 4346: Part 3. Any cleaning fluid used should be from the same manufacturer as the solvent cement.

Figures 11.29 and 11.30 show examples of typical solvent cement joints.

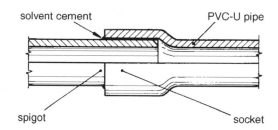

solvent cement   PVC-U pipe

spigot   socket

Jointing method:

(1) Cut pipe end square and remove internal burrs.
(2) Slightly chamfer outer edge of pipe (at about 15° to pipe axis).
(3) Roughen joint surfaces, using clean emery cloth or medium glasspaper.
(4) Degrease joint surfaces of pipe and fitting with cleaning fluid, using absorbent paper.
(5) Using a brush, apply an even layer of cement to both fitting and pipe in a lengthwise direction, with a thicker coating on the pipe.
(6) Immediately push the fitting on to the pipe without turning it. Hold for a few seconds, then remove surplus cement.
(7) Leave undisturbed for five minutes, then handle with reasonable care.
(8) Allow 8 hours before applying the full rated pressure, and 24 hours before testing at one-and-a-half times the full rated pressure.
    For lower pressure, allow one hour per bar, e.g. 3 bar would require 3 hours drying time.

Figure 11.29   Solvent cement joint for PVC-U pipe

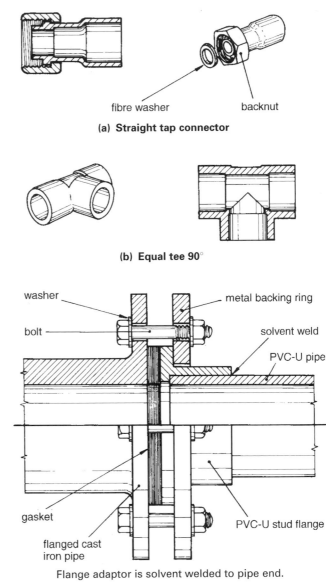

fibre washer   backnut

(a) Straight tap connector

(b) Equal tee 90°

washer   metal backing ring

bolt   solvent weld

PVC-U pipe

gasket

PVC-U stud flange

flanged cast
iron pipe

Flange adaptor is solvent welded to pipe end.

(c) Flanged connections for PVC-U

Figure 11.30   Examples of solvent cement joints

Chlorinated polyvinyl chloride pipes (PVC-C)

PVC-C pipes to BS 7291: Parts 1 and 4 are available in sizes 12 mm to 63 mm and are suitable for both hot and cold water applications. Jointing is similar to that shown for PVC-U. See table 11.13 for fixing distances.

Cross-linked polyethylene (PE-X)

Pipes and fittings of cross-linked polyethylene (PE-X) to BS 7291: Parts 1 and 3 are suited to both hot and cold water applications. Joints may be of the push-fit or compression types and will be similar to those illustrated for polyethylene.

Polybutylene (PB)

Polybutylene should conform to BS 7291: Parts 1 and 2 and can be used for both hot and cold water. It is also suited to use in cold temperature installations. Fittings may be of the compression or push-fit types.

Propylene copolymer (PP)

Propylene copolymer pipes and fittings to BS 4991 are suited only to cold water applications up to 20°C. Above these temperatures the material's properties are reduced.

Table 11.14 gives further general information on these materials and spacing for fixings is shown in table 11.15.

Table 11.14   Plastics pipes and their uses

Pipe material	Abbreviation	BS number	Suitable for hot or cold	Temperature limits	Range of pipe sizes	Jointing methods
Polyethylene (blue)	PE	BS 6572	Cold For below ground use only	20°C max.	20 to 63	Compression Push-fit Thermal fusion
Polyethylene (black)	PE	BS 6730	Cold For above ground use only	20°C max.	20 to 63	Compression Push-fit Thermal fusion
Polyethylene (Type 32)	PE	BS 1972	Cold For above ground use only	20°C max.	10 to 35	Compression Push-fit Thermal fusion
Unplacticised polyvinyl chloride	PVC-U	BS 3505 BS 3506	Cold Non-drinking	20°C max. 20°C max.	$\frac{3}{8}''$ to 24'' $\frac{3}{8}''$ to 24''	Compression Push-fit Solvent cement weld Mechanical
Propylene copolymer	PP	BS 4991	Cold	20°C max.	$\frac{1}{4}''$ to 24''	Compression Push-fit
Polybutylene	PB	BS 7291: Parts 1 and 2	Cold and hot	83°C max.	10 to 35 and 10 to 32	Compression Push-fit
Cross-linked polyethylene	PE-X	BS 7291; Parts 1 and 3	Cold and hot	95°C max.	10 to 35 and 10 to 32	Compression Push-fit
Chlorinated polyvinyl chloride	PVC-C	BS 7291; Parts 1 and 4	Cold and hot	83°C max.	12 to 63	Compression Push-fit Solvent cement weld

Table 11.15   Maximum spacing of fixings for internal PB and PE-X pipes

Type of piping	Nominal size of pipe mm	Spacing on horizontal run mm	Spacing on vertical run mm
Polybutylene (PB) to BS 7291: Parts 1 and 2 [and] cross-linked polyethylene (PE-X) to BS 7291: Parts 1 and 4	up to 16 18 to 25 28 32 35	300 500 800 900 900	500 800 1000 1200 1200

11.5   Iron pipes

These are made in three types: vertically cast, spun iron and ductile iron. However, the production of vertically cast and spun iron pipes has now virtually ceased in favour of pipes in ductile iron to BS 4772, which have much improved mechanical properties.

Since all iron pipes are liable to corrosion they are factory treated inside and out. Also, many water authorities may require pipes to be sheathed in a blue polyethylene sleeve to BS 6076 for further protection from aggressive soils, and for identification.

Ductile iron pipes up to 1600 mm in diameter are considered suitable for working pressures of up to 40 bar depending on the size and pressure rating, and should be pressure tested to 5 bar above the expected working pressure.

Fixings for cast iron pipes

See figure 11.31 and table 11.16 for examples of fixings and spacings for cast iron pipes.

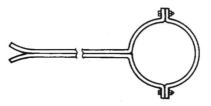

(a) Holderbat build-in type in mild steel

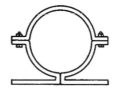

(b) Holderbat screw-to-wall type in mild steel

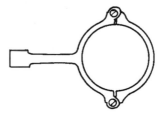

(c) Holderbat hinged build-in type in cast iron

Figure 11.31   Brackets for cast iron pipe

Table 11.16   Maximum spacing of fixings for above ground cast iron pipes

Type of piping	Nominal size of pipe mm	Spacing on horizontal run m	Spacing on vertical run m
Vertically cast or spun iron complying with BS 1211 or BS 2035	51 76 102 152	1.8 2.7 2.7 3.6	1.8 2.7 2.7 3.6
Ductile iron complying with BS EN 545, BS EN 598	80 100 150	2.7 2.7 2.7	2.7 2.7 3.6

Jointing methods for cast iron pipes

See figures 11.32–11.37.

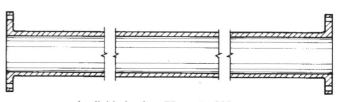

Available in sizes 75 mm to 300 mm.

Figure 11.32   **Flanged spun iron and cast iron pipes**

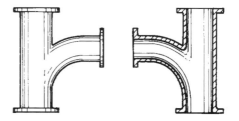

Figure 11.33   **Flanged junction**

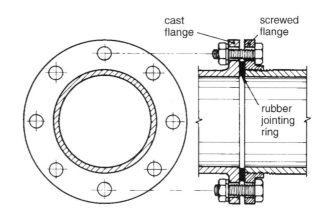

Figure 11.34   **Flanged joint detail**

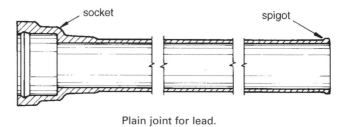

Plain joint for lead.

Figure 11.35   **Cast iron pipe with plain socketed joint**

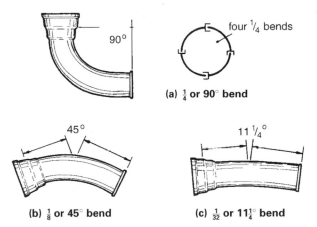

(a) $\frac{1}{4}$ or **90° bend**

(b) $\frac{1}{8}$ or **45° bend**

(c) $\frac{1}{32}$ or **11$\frac{1}{4}$° bend**

Figure 11.36   Plain socketed pipe bends

Caulking tool chosen to match pipe size and width of joint.

Caulk when cold to finish about 3 mm inside socket face.

Use synthetic yarn that will not promote the growth of bacteria. Yarn should be caulked tightly to approximately one-third depth of the joint, to prevent direct contact between lead and water, and to centralize pipe in socket.

No longer permitted for use on potable water installations.

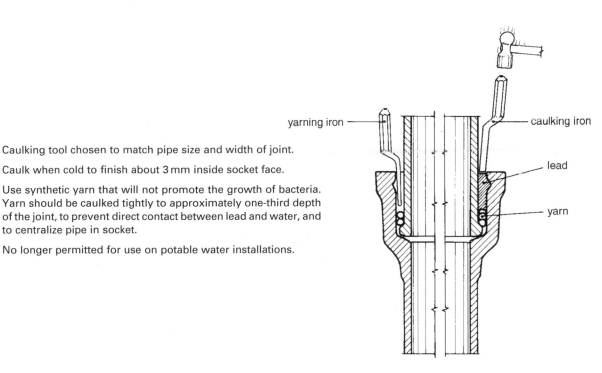

Figure 11.37   Lead run joint for cast iron pipes

Ductile iron pipe will use flanged joints or more commonly push-fit 'O' ring type mechanical joints (see figures 11.38 and 11.39, giving good flexibility of movement without loss of joint seal.

Other joints used for cast iron pipes are the Viking Johnson (see figure 11.40) and the bolted gland joint (see figure 11.41).

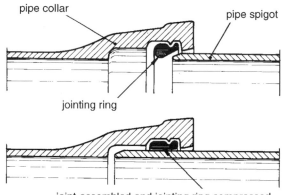

Available in sizes 80 mm to 600 mm.

Suitable for spun iron and ductile iron pipes

Figure 11.38   **'Tyton' slip-fit joint**

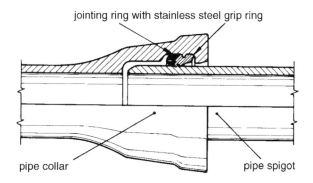

Stainless steel grip ring makes joint resistant to pulling apart.

Figure 11.39   **Slip-fit joint with self-anchoring gasket**

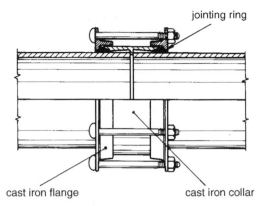

cast iron flange   cast iron collar

Also suitable for PVC-U, steel and asbestos cement pipes.

(a) Straight coupler

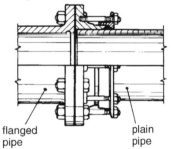

flanged   plain
pipe   pipe

Used for jointing plain ended pipes to flanged pipes or fittings.

Each adaptor consists of flanged sleeve, end flange, wedge rubber packing ring, bolts, studs and nuts.

(b) Flange adaptor

Figure 11.40   Viking Johnson joints

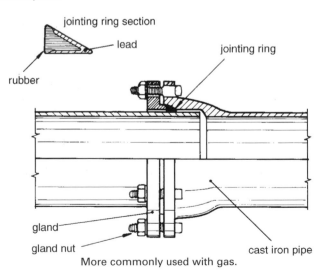

More commonly used with gas.

Figure 11.41   Bolted gland joint

11.6 Asbestos cement pipes

Asbestos cement pipes, in common with many asbestos-based materials, are subject to the requirements of the Asbestos Regulations 1969 and the Health and Safety at Work etc. Act 1974. These regulations require that any dust liberated is restricted to a low level of concentration. Asbestos cement pipes contain only a small percentage of asbestos and are considered safe to handle. Where a limited amount of cutting and turning by hand tools is done in the open air, the dust level is generally below the minimum set in the regulations. However, if any doubt exists, clarification must be sought from the local office of HM Factory Inspectorate, or from the manufacturer of the material.

Asbestos cement pipe is corrosion-resistant in most soils but since it is very brittle it is liable to break where soils, such as clays, move because of seasonal loss or gain of moisture.

The use of asbestos cement pipes is restricted to installations below ground. Pipe is available from 50 mm to 900 mm diameter. See table 11.17 and figures 11.42 and 11.43.

Table 11.17   **Class and pressure range of asbestos cement pipes**

Class	Metric colour coding (pipes, joints and rings)	Works test pressure bar	Maximum working pressure bar	Test pressure
15	Green	15	7.5	One-and-a-half to twice the expected working pressure
20	Blue	20	10	
25	Violet	25	12.5	

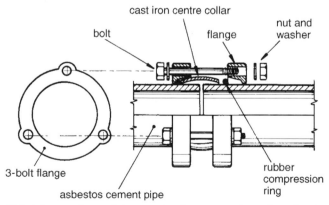

This joint needs careful protection from corrosion otherwise the advantage of the non-corrosive properties of asbestos cement pipes is lost.

Made also for cast iron pipes.

Figure 11.42   **Detachable joint made of cast iron**

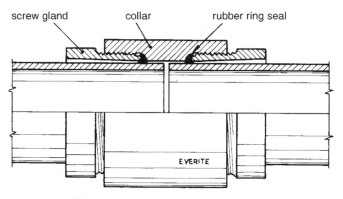

screw gland   collar   rubber ring seal

EVERITE

Gives complete immunity from corrosion.

Figure 11.43   All asbestos cement screwed gland joint

11.7   Lead pipes

No lead pipe or other water fitting containing lead is permitted to be used in water systems containing wholesome water, even for repair. Where new pipes, e.g. copper, are to be connected to an existing lead pipe, protection against electrolytic (galvanic) action is required. If possible lead pipes should be removed.

Lead solders are not permitted for use on pipelines for wholesome water.

11.8   Connections between pipes of different materials

As far as possible connections between different pipe materials should be avoided especially when jointing dissimilar metals, as this may lead to electrolytic corrosion. However, when dissimilar pipes have to be used, for example, in the repair or renewal of part systems or connections to existing pipelines, they may be connected by one of the following three methods:

(1)   using 'inert' material between the dissimilar metals (see figure 11.44);
(2)   ensuring that water flows towards a potentially stronger material in the electrochemical series from a weaker one:

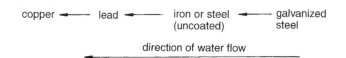

copper ◄──── lead ◄──── iron or steel ◄──── galvanized
                        (uncoated)   steel

direction of water flow

Water flowing in the opposite direction will carry dissolved particles of potentially stronger material which will adversely affect the weaker pipeline;
(3)   using a sacrificial anode of a potentially weak material (see figure 11.45).

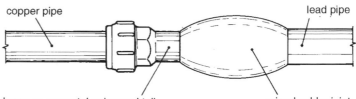

direction of flow

copper pipe

lead pipe

brass or gunmetal union and tail

wiped solder joint

Brass or gunmetal tail used to prevent direct contact between copper pipe and lead pipe. Direct contact is not permitted.

The wiped soldered joint is the traditional means of connecting to existing lead pipes.

An alternative jointing method is to use a 'patent' compression joint.

Figure 11.44   Use of inert material between connections of dissimilar metals

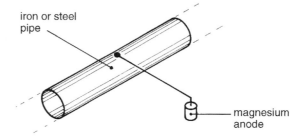

iron or steel pipe

magnesium anode

Anode will corrode and leave the pipe intact.

Sacrificial anodes can be fitted to cisterns, tanks and cylinders.

Figure 11.45   Sacrificial anode used on a pipeline

Adaptor couplings are available for the joining of a wide range of differing materials. A number of these have been shown in previous illustrations.

11.9   Connections to cisterns and tanks

The following general instructions should be followed.

(1)   Provide proper support for cisterns and tanks to avoid undue stress on connections and deformation of cistern or tank walls.
(2)   Use proper tools for hole cutting (see figure 11.46) not flame cutters, not hammer and chisel.
(3)   Holes must be truly circular with clean edges.
(4)   All debris or filings must be removed from inside the tank or cistern.

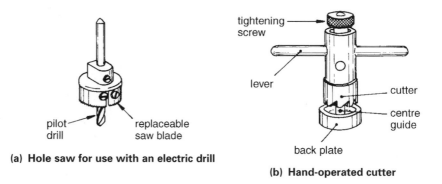

(a) **Hole saw for use with an electric drill**

(b) **Hand-operated cutter**

Figure 11.46   **Tools for cutting holes.**

Figure 11.47 shows examples of cistern connections.
Additional considerations for thermoplastics cisterns:

(1) Scribing tools should *not* be used to mark position of holes.
(2) Cistern wall should be supported with wooden strut or similar during cutting.
(3) Pipes should be carefully fitted and supported to avoid distortion of cistern or tank.
(4) Corrosion-resisting support washers should be used inside and outside the cistern to strengthen the joint area.
(5) Float valves should be fitted through a supporting back plate to stabilize the cistern wall against the thrust of the lever arm.
(6) Linseed oil based sealants must not be used with plastic cisterns or pipes. Use only PTFE sealants.

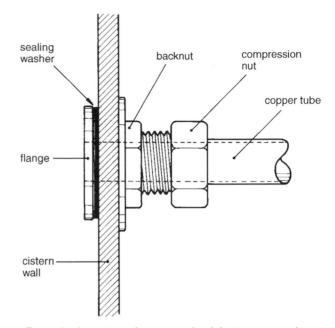

Example shows type A compression joint to copper tube

(a) Copper connection to cistern

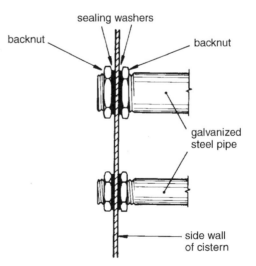

Copper pipe should not be connected to galvanized steel cisterns or tanks.

Pipe sealed with proprietary washers.

(b) Galvanized steel connections to steel cistern

Figure 11.47   Examples of cistern connections

continued

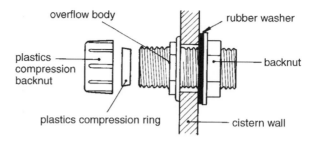

(c) Plastics overflow connection to WC cistern

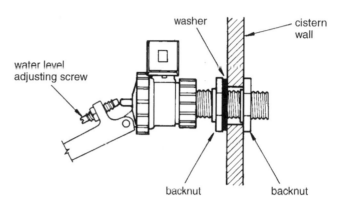

(d) Plastics float valve connection to cistern

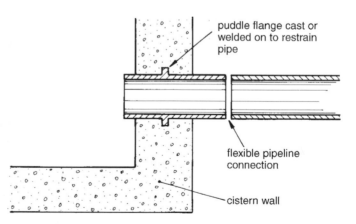

Puddle flange properly aligned before casting into concrete.

Concrete to be compacted to ensure watertight joint.

(e) Connection to concrete tank or cistern

Figure 11.47 **Examples of cistern connections**
continued

11.10 Branch connections for buildings

Service connections to mains are normally made by drilling and tapping the top of the main and screwing in a union ferrule (see figures 11.48 and 11.49).

The ferrule is set on top of the main to avoid drawing sediment into the service pipe.

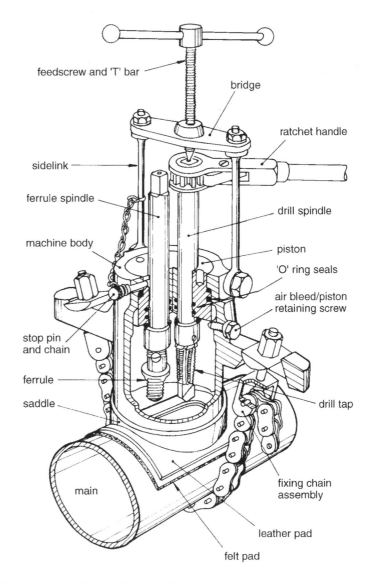

Figure 11.48   **Under pressure mains tapping machine**

Ferrule screwed directly into top of cast iron water main.

Branch pipe to run parallel to main before taking its final route.

Gooseneck to permit movement in soil without damage to pipe.

Gooseneck bent to ensure that ferrule will not loosen if the service pipe should settle.

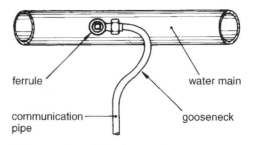

ferrule   water main

communication   gooseneck
pipe

(a) Connection of ferrule to main

Mains pipes of asbestos cement, steel and PVC-U are strengthened at the point of connection by the use of a strap saddle.

Strap saddle of cast iron for asbestos cement and steel mains, and of gunmetal for PVC-U mains.

main of steel,   communication
asbestos cement   pipe
or PVC-U

                                                   ferrule

strap saddle

(b) Use of strap saddle for mains connections

Self-tapping ferrule is shown here.

Strap saddles of PVC-U are available for solvent welding or with mechanical connection to the main.

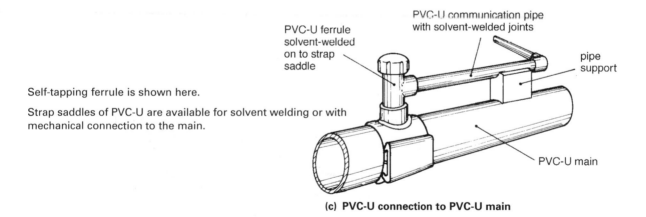

PVC-U ferrule
solvent-welded
on to strap
saddle

PVC-U communication pipe
with solvent-welded joints

pipe
support

PVC-U main

(c) PVC-U connection to PVC-U main

Figure 11.49   **Mains connection**

Where rigid pipes are used for the branch service pipe connect them using a short length of suitable flexible pipe.

Depending on the size of the service pipe and the main, ferrules may not always be suitable. For larger connections or smaller mains a more suitable method would be the use of a leadless collar or a tee connection (see figures 11.50 and 11.51, and table 11.18).

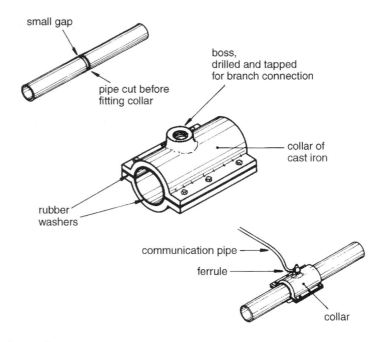

Figure 11.50   Leadless collar

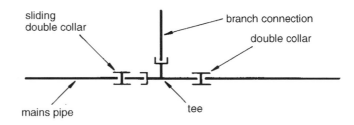

Figure 11.51   Tee connection

Table 11.18   **Method of branch pipe connection**

Nominal size of branch pipes		Nominal diameter of main pipe				
mm	inches	80 mm	100 mm	150 mm	200 mm	250 mm and over
15	$\frac{1}{2}$	F	F	F	F	F
22	$\frac{3}{4}$	T	F	F	F	F
25	1	T	T	F	F	F
35	$1\frac{1}{4}$	T	T	T	F	F
42	$1\frac{1}{2}$	T	T	T	F	F
54	2	T	T	T	T	F

F: Ferrule   T: Tee or leadless collar

11.11   Contamination of mains

When cutting into mains to make branch connections, precautions should be taken to avoid contamination.

(1)   Sterilize the trench around the branch connection before cutting.
(2)   Take care to avoid the entry of soil or water from the trench.
(3)   Insert sterilizing tablets into the pipe when making the connection.

11.12   Laying underground pipes

Pipes should be laid on a firm even base, evenly supported throughout their length, and must not rest on their sockets, bricks or other makeshift supports. Plastics pipes should be laid on a bed free from sharp stones.

As far as possible pipes should be laid in straight lines to permit easy location later. However, copper or plastics pipes should be snaked within the trench to allow for settlement and moisture movement in the soil (see figure 11.52).

Joints below ground, and in other inaccessible places, should be kept to a minimum and avoided where possible. It is preferable that pipes are laid in one length, e.g. copper and polyethylene pipes.

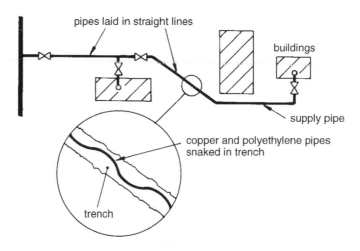

Large diameter PVC-U pipes must not be bent within their length. However, some deflection is permissible at flexible joints.

Figure 11.52   **Location of pipelines**

Trench excavations

The bottoms of trenches must be carefully prepared to a firm even surface so that the barrels of the pipes when laid are well bedded down for their whole length (see figure 11.53). Mud, rock projections, boulders, hard spots and local soft spots should be removed and replaced with selected fill material consolidated to the required level. The width of trenches must be sufficient to allow the pipes to be properly bedded, jointed and backfilled.

Where rock is encountered, the trench should be cut at least 150 mm deeper than other ground and made up with well rammed bedding material.

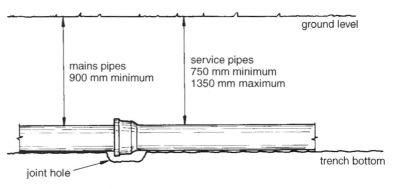

ground level

mains pipes
900 mm minimum

service pipes
750 mm minimum
1350 mm maximum

trench bottom

joint hole

Pipeline to follow contours of ground level whilst ensuring adequate depth of cover.

Pipe to lie flat on trench bottom if soil conditions permit.

Additional excavation needed to leave space for collar.

Figure 11.53   Support of pipes in trench

Pipes must be kept clean and each pipe and fitting must be thoroughly cleansed internally immediately before fitting. The open end should be temporarily plugged until jointing takes place. Particular care must be taken to keep the joints clean. After laying and jointing, the leading end must remain plugged.

Precautions must be taken to prevent the plugged pipes floating if the trench becomes flooded.

Coatings, sheathings or wrappings must be examined for damage, and repaired where necessary, before trenches are backfilled.

When backfilling trenches (see figure 11.54), the pipes must be surrounded by a selected material consolidated to prevent subsequent pipe movement. No large stones or sharp objects should be in contact with the pipes.

PVC-U pipes need extra care, especially when being backfilled, otherwise pipes will become distorted and weakened. Granular bedding and surrounds are essential (see figure 11.55).

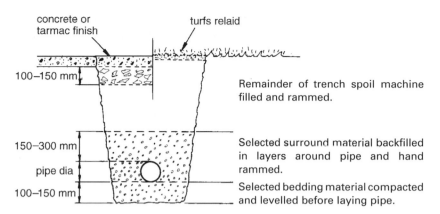

concrete or
tarmac finish

turfs relaid

100–150 mm

Remainder of trench spoil machine
filled and rammed.

150–300 mm

pipe dia

Selected surround material backfilled
in layers around pipe and hand
rammed.

100–150 mm

Selected bedding material compacted
and levelled before laying pipe.

Reinstatement depends on surrounding surface, e.g. grass verge or road.

Figure 11.54   Trench backfilling

Light compacting by machine

Compact by hand in layers

3608 granular surround will give
all-round protection and prevent
ovalling of pipe.

300 mm

pipe dia

100–150 mm

Figure 11.55   Backfilling trenches for PVC-U and PE pipes

Restraint of pipes

With most methods of pipe laying and trench backfill for large diameter
pipes, joints at changes in direction are liable to move and push apart due
to internal thrust pressure (see figure 11.56). To guard against these risks,
which will vary according to the internal pressure and how much the pipe
direction changes, pipes must be securely anchored at bends, branches and
pipe ends. The amount of anchorage required will also depend upon the
soil and its bearing capacity. See also tables 11.19–11.21.

Calculation of the thrusts which act in the direction of arrows shown in
figure 11.56 may be calculated using the formulae below.

- end thrust (kN) in capped ends and branches = $100\,AP$
- radial thrust at bends (kN) $= 100\,AP \times 25\text{in}\theta/2$

where
A is the cross sectional area of the socket (m^2)
P is the test pressure (bar)
θ is the angle of deviation of the bend.

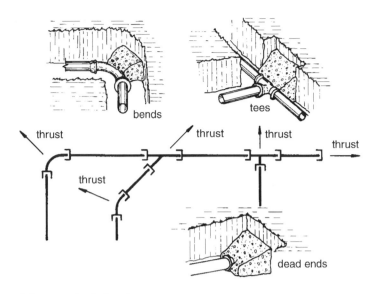

Figure 11.56   **Horizontal thrust on buried mains**

Alternatively, when standard fittings are used, the thrusts may be calculated by multiplying the values given in table 11.19 by the test pressure in bars.

Thrust blocks used to restrain pipelines must have sufficient bearing area to resist the thrust under test pressure. This can be calculated by using data for soil-bearing capacities given in table 11.21.

Table 11.19   **Thrust per bar internal pressure**

Nominal internal diameter of pipe	End thrust	Radial thrust on bends of angle			
		$90°$	$45°$	$22\frac{1}{2}°$	$11\frac{1}{4}°$
mm	kN	kN			
50	0.38	0.53	0.29	0.15	0.07
75	0.72	1.02	0.55	0.28	0.15
100	1.17	1.66	0.90	0.46	0.24
125	1.76	2.49	1.35	0.69	0.35
150	2.47	3.50	1.89	0.96	0.49
175	3.29	4.66	2.52	1.29	0.65
200	4.24	5.99	3.24	1.66	0.84
225	5.27	7.46	4.04	2.06	1.04
250	6.43	9.09	4.92	2.51	1.26
300	9.38	13.26	7.18	3.66	1.84
350	12.53	17.71	9.59	4.89	2.46

Table 11.20   **Gradient thrusts on buried or exposed mains**

Gradient	Spacing of anchor block	
	m	ft
1 in 2	5.5	18
1 in 3	11.0	36
1 in 4	11.0	36
1 in 5	16.5	54
1 in 6	22.0	72

gradient 1 in 6
or steeper

Table 11.21   **Bearing capacity of soils**

Soil type	Safe bearing load kN/m^2
Soft clay	24
Sand	48
Sandstone and gravel	72
Sand and gravel bonded with clay	96
Shale	240

Valve chambers and surface boxes

Surface boxes must be provided to allow access to valves and hydrants, and must be supported on concrete on brickwork which, after making an allowance for settlement, must not rest on the pipes and transmit loads to them. See figure 11.57.

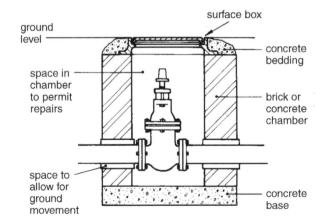

Figure 11.57   **Valve chamber**

Brick or concrete hydrant chambers (see figure 11.58) must be large enough to permit repairs to be carried out to the fittings.

An alternative is to provide vertical guard pipes or precast concrete sections to enclose the spindles of valves as shown in figure 11.59.

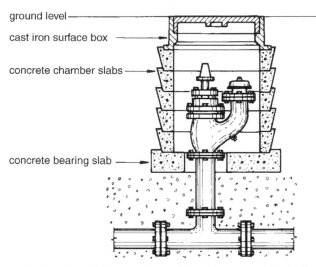

ground level

cast iron surface box

concrete chamber slabs

concrete bearing slab

Concrete chamber slabs provide quick and easy method of building chamber plus easy removal for access for repairs.

Hydrant installations to comply wtih BS 3251.

Internal dimensions 430 mm × 280 mm.

Figure 11.58   Typical hydrant arrangement

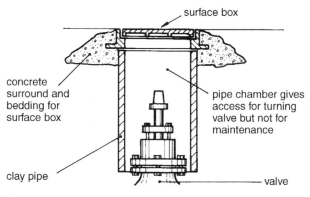

surface box

concrete surround and bedding for surface box

pipe chamber gives access for turning valve but not for maintenance

clay pipe

valve

Commonly used because it is cheap, but valve must be dug out for repairs.

Figure 11.59   Valve access

All valves and hydrants should be positioned where they can easily be found and used.

Surface boxes must be positioned and marked to indicate the pipe service, the size of mains, and position and depth below the surface (see figure 11.60).

Surface boxes should be of sufficient strength to withstand likely traffic loads, i.e.

BS 750 Class A for use in carriageways where loads may be heavy.

Class B for use where vehicles would have only occasional access such as footpaths and verges.

Indicator plates, illustrated in figure 11.61, can be screwed to walls or marker posts, but must be visible.

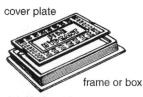

(a) Hydrant box

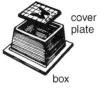

(b) Valve box

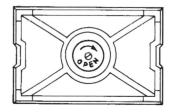

(c) Hydrant cover marking

Cover plates marked to identify type of valve enclosed in chamber.

Hydrant cover must be marked underneath to indicate direction of valve turning.

Figure 11.60 Surface boxes for valves and hydrants

Indicator plates for hydrants should conform to BS 3251

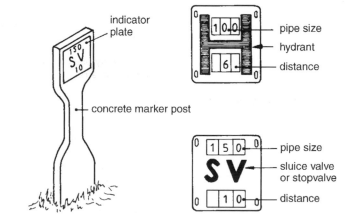

Figure 11.61 Valve and hydrant marker plates

Drawings should show all pipe runs, valves and hydrants. Working drawings should be amended to show variations from the original design.

11.13   Pipework in buildings

Fixings and allowance for thermal movement

Allowance must be made for expansion and contraction of pipes (see figure 11.62), especially where pipe runs are long and where temperature changes are considerable (hot distributing pipes) and where the pipe material has a relatively high coefficient of thermal expansion, e.g. plastics.

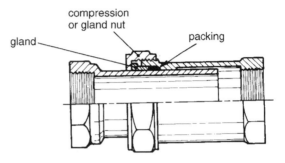

(a) Gland type expansion joint

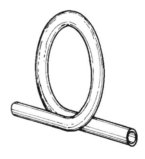

(b) Expansion loop

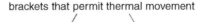
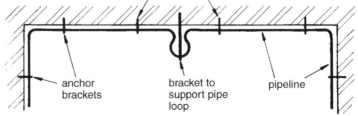

(c) Use of horseshoe expansion loop

Figure 11.62   **Expansion joints**

Fixing insulated piping

Sufficient space should be allowed behind pipes for insulation to be properly installed, as shown in figure 11.63.

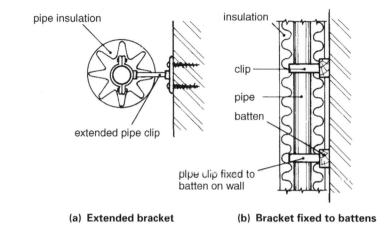

(a) Extended bracket   **(b) Bracket fixed to battens**

Figure 11.63   **Fixings for insulated piping**

Concealed pipes

These must be housed in properly constructed ducts or chases that provide access for maintenance and inspection. See chapter 10.

Pipes passing through structural timbers

These must not be weakened by indiscriminate notching and boring. Notches and holes should be as small as practicable. Whenever possible, notches should be U-shaped and formed by parallel cuts to previously bored holes (see figure 11.64). The positions of notches and holes are as important as their sizes, and in timber beams and joists should only be made within the hatched areas shown in figure 11.65.

Before making notches on holes, agreement should be reached on their size and position with the architect, structural engineer or building supervisor.

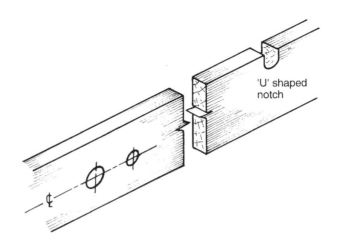

'U' shaped notch

Holes and notches to be as small as practicable but large enough to permit pipes to expand and contract.

To avoid weakening the joist, holes and notches should be restricted to the hatched areas shown in figure 11.65.

Figure 11.64   Pipes passing through structural timbers

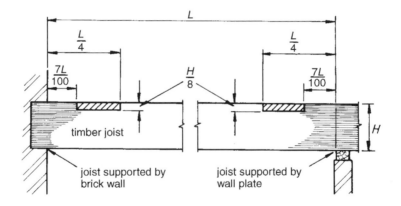

(a) Notches

Note Notches must be restricted to the hatched areas shown if the joists are not to be weakened.

L is the length of the structural member
H is the height of the structural member
If H exceeds 250 mm then for calculation purposes it is deemed to be 250 mm

Example For a joist 6 m long and 300 mm deep, notches must be:

(a) not more than H/8 deep, i.e.:

$$\frac{250}{8} = 31 \text{ mm}$$

(b) (i) at least 7 L/100 from bearing, i.e.:

$$\frac{7 \times 6000}{100} = \frac{42000}{100} = 420 \text{ mm}$$

(ii) not more than L/4 from bearing, i.e.:

$$\frac{6000}{4} = 1500 \text{ mm} = 1.5 \text{ m}$$

Figure 11.65   Dimensions for notches and holes

continued

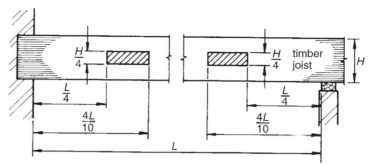

(b) Holes

Note Holes must be restricted to the hatched areas shown if the joists are not to be weakened.

L is the length of the structural member

H is the height of the structural member

If *H* exceeds 250 mm then use *H* = 250 mm for calculation purposes.

Example For a joist 4 m long and 200 mm deep, holes must be:

(a) not more than *H*/4 in diameter, i.e.

$$\frac{200}{4} = 50 \text{ mm}$$

(b) (i) at least *L*/4 from bearing, i.e.

$$\frac{4000}{4} = 1000 \text{ mm} = 1 \text{ m}$$

(ii) not more than 4*L*/10 from bearing, i.e.

$$\frac{4 \times 4000}{10} = 1600 = 1.6 \text{ m}$$

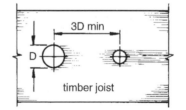

(c) Spacing of holes

The distance between adjacent holes in joists should be at least three times the diameter *D* of the larger hole.

Holes should pass through the centre of the joist and be parallel.

Figure 11.65 continued   **Dimensions for notches and holes**

Penetration of fire walls and floors

As required by the current Building Regulations, penetration of compartment walls and floors and fire barriers must be fire-stopped to prevent the passage of smoke and flame.

Only pipes of non-combustible materials such as cast iron, steel or copper are permitted to pass through compartment walls; a non-combustible material being one that will not soften or melt when exposed to a temperature of 800°C. Additionally, these pipes must be no greater in diameter than 160 mm.

Where pipes of PVC or other plastics are taken through compartment walls they must pass through a sleeve of non-combustible material.

11.14 Electrical earthing and bonding

Water pipes should not be used as an electrode for earthing purposes, but all metal pipes entering buildings should be bonded to the electrical installation as near as possible to the point of entry to the building.

Where pipes, fitting or appliances are to be replaced, the earth continuity and equipotential bonding should be maintained. Meter installations, for example, that are designed to permit replacement, should have a suitable conductor permanently fitted between inlet and outlet pipework. Where no permanent conductor is in place a temporary conductor should be fitted for the duration of the replacement work. Any electrical earthing or bonding installation that has been disturbed should be properly tested to ensure continuity is maintained.

Electrical installations, including earthing and bonding arrangements, should be carried out by a suitably qualified electrical contractor enrolled with the National Inspection Council for Electrical Installation Contracting (NICEIC). Electrical installations should comply with BS 7671 (Requirements for Electrical Installations – IEE Wiring Regulations).

11.15 Jointing of pipework for potable water

Research, and the testing of water samples from mains and services, have shown that some traditional jointing materials harbour or promote the growth of bacteria. As a result, some of these materials are now banned and others discouraged.

For example, bacterial contamination may be caused by linseed oil-based compounds commonly used with screwed joints. New compounds now available do not promote bacterial growth.

Again, if soldered joints are badly made, lead can be leached into solution and consumed. So, the Water Regulations now permit only lead-free solder in capillary joints for hot and cold water supplies.

However, there are problems when choosing jointing materials because for some applications no suitable alternatives are available for the traditional materials.

Table 11.22 adapted from BS 6700, lists permitted jointing materials and gives guidance on their use.

Table 11.22   Jointing of potable water pipework

This table lists jointing methods and materials that should be used for jointing potable water pipework. Materials and products that have been assessed under the Water Regulations Advisory Scheme and listed by them are considered to meet the requirements of this table.

Type of joint	Method of connection	Jointing material	Precautions/limitations
Lead to brass, gunmetal, or copper, pipe or fitting	Plumber's wiped soldered joint	Tallow Flux	Of very limited application See note
Copper to copper pipe (pipe to pipe or fitting, or fitting to fitting)	Capillary-ring or end feed, soldered joint	Flux	No solder containing lead to be used See note
Copper to copper (pipe to fitting or fitting to fitting)	Non-manipulative compression fittings	—	Above ground only
	Manipulative compression fittings	Lubricant on pipe end when required	See note
Copper to copper (pipe to fitting or fitting to fitting)	Bronze welding or hard solder	Flux	See note
Galvanized steel (pipe to pipe or fitting), including copper alloy fittings	Screwed joint, where seal is made on the threads	PTFE tape or proprietary sealants	PTFE tape only up to 40 mm ($1\frac{1}{2}$ in diameter See note
Galvanized or copper (pipe to pipe or fitting)	Flanges	Elastomeric joint rings complying with BS 2494, or corrugated metal. Vulcanized fibre rings complying with BS 5292 or BS 6091: Parts 1 and 2	See note
Long screw connector	Screwed pipework with BS 2779 thread	Grummet made of proprietary paste and hemp	Must not promote growth of bacteria
Shouldered screw connector	Seal made on shoulder with BS 2779 thread	Elastomeric joint rings complying with BS 2494 and plastics materials	—
Unplasticized PVC (pipe to fitting)	Solvent welded in sockets	Solvent cement complying with BS 4346: Part 3	Follow manufacturers' recommendations
	Spigot and socket with ring seal. Flanges. Union connectors	Elastomeric seal complying with BS 2494. Lubricants	Lubricant should be compatible with the unplasticized PVC and elastomeric seal
Cast iron (pipe to fitting)	Caulked lead	Sterilized gaskin yarn/blue lead	See note
	Bolted or screwed gland joints	Elastomeric ring complying with BS 2494	—

continued

Table 11.22   **continued**

Type of joint	Method of connection	Jointing material	Precautions/limitations
	Spigot and socket with ring seal	Elastomeric seal and lubricant	—
Copper or plastic (pipe to tap or float-operated valve)	Union connector	Elastomeric or fibre washer	—
Stainless steel (pipe to fitting, including copper alloy fittings	Non-manipulative compression fittings	Lubricant when required	See note
	Manipulative fittings	Elastomeric seals when required	—
Pipework connections to storage cisterns (galvanized steel, reinforced plastics, polypropylene, polyethylene)	Tank connector/union with flanger backnut	Washers: elastomeric, polyethylene, fibre	—
Polyethylene (pipe to fitting)	Non-manipulative fittings	—	Do not use lubricant
	Thermal fusion fittings	—	Follow manufacturers' directions
Polybutylene (pipe to pipe or fitting)	Non-manipulative fittings	Lubricant on pipe end when required	Lubricant if used should be listed and compatible with plastics
	Thermal fusion fittings	—	Follow manufacturers' directions
Polypropylene (pipe to pipe)	Non-manipulative fittings	Lubricant on pipe end when required	Lubricant if used should be listed and compatible with plastics
	Thermal fusion fittings	—	Follow manufacturers' directions
Cross-linked polyethylene (pipe to fitting)	Not-manipulative fittings	Lubricant on pipe end when required	Lubricant if used should be listed and compatible with plastics
Chlorinated PVC (pipe to fitting)	Solvent welded in sockets	Solvent cement to BS 4346: Part 3	Follow manufacturers' recommendations

Note. Where non-listed materials are to be used, due to there being no alternative, the procedure used should be consistent with the manufacturer's instructions taking particular note of the following precautions:

   (1) use least quantity of material to produce good quality joints;
   (2) keep jointing materials clean and free from contamination;
   (3) remove cutting oils and protective coatings, and clean surfaces;
   (4) prevent entry of surplus materials to waterways;
   (5) remove excess materials on completion of the joint.

11.16   Testing

Smaller services should be tested by filling up the system at normal working pressure and inspecting all joints and fittings for leakage.

Any pipes below ground, buried under screeds, or in other inaccessible places should be tested before being covered. Hot water pipes should be similarly checked after heat has been applied. See chapter 12 for further information on testing.

11.17   Identification of valves and pipes

Below ground

The position of underground pipes and the location of valves should be recorded on a plan of the premises. Valve surface boxes should be marked to indicate the service below them. On mains and larger service pipes, indicator plates should be set up to show the size and position of valves and hydrants. These are illustrated in figures 11.60 and 11.61.

Water pipes below ground should be coloured blue to distinguish them from other services.

Above ground

Valves on hot and cold pipes should be fitted with an identification label of non-corrodible and non-combustible material, as shown in figure 11.66(a). The label should describe the size and function of the valve.

Alternatively, the label may be marked with a reference number which relates to a durable diagram of the water system showing valve reference numbers, and fixed in a prominent position. See figure 11.66.

Pipes within buildings should be colour banded as shown in figure 11.67, to identify the pipe and the service for which it is used.

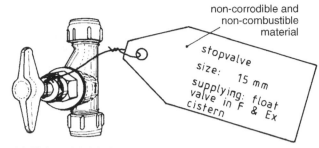

(a) Valve with label

Key

Control valve	Pipe run	
SV1	sp3	
SV2	sp2	supply pipe
SV3	sp4	
SV4 Servicing	h1	distributing header
SV5 valves	h1	
SV6	dp1	distributing pipe
SV7	cf1	cold feed pipe
SV8	dp2	distributing pipe

Cisterns	
CWSC No. 1	Combined feed and storage
CWSC No. 2	cistern (linked)
F & Ex C	Feed and expansion cistern

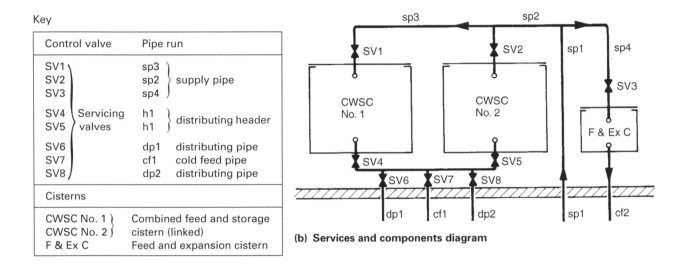

(b) Services and components diagram

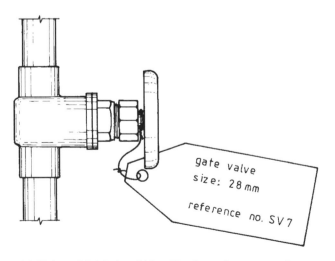

(c) Valve with label and identification reference number

Figure 11.66 **Identification of pipes and services**

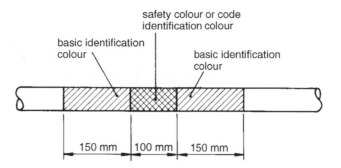

safety colour or code
identification colour

basic identification
colour

basic identification
colour

150 mm | 100 mm | 150 mm

Colour code identification may be one or more colours. Dimensions show minimim length for each colour.

(a) Colour coding

pipe contents	basic colour	colour code indication			basic colour
drinking water	green	auxiliary blue			green
boiler feed	green	crimson	white	crimson	green
central heating	green	blue	crimson	blue	green
cold down service	green	white	blue	white	green
hot water supply	green	white	crimson	white	green
fire extinguishing	green	red			green

(b) Example of colours used for water pipelines

Figure 11.67   Identification of pipelines above ground to BS 1710

Chapter 12
Commissioning and maintenance of pipelines, services and installations

Commissioning includes inspection, testing and cleansing of pipes and installations, and is an integral part of the work. It should be allowed for when estimating costs and should be undertaken at appropriate times as the work proceeds.

Inspection, testing and cleansing should ensure that:

(1) materials and equipment conform to British Standards or other forms of approval;
(2) installations are in accordance with the specification;
(3) all relevant laws and regulations are complied with;
(4) water drawn from pipes and fittings is of good quality and fit for human consumption.

12.1 Inspections

Inspection of below ground installations

During the visual inspection of pipelines, particular attention must be paid to the pipe bed, the line and level of the pipe, irregularities at joints, the correct fitting of air valves, washout valves, sluice valves and other valves together with any other mains equipment specified. This should include the correct installation of thrust blocks, where required, and ensure that protective coatings are undamaged.

Trenches must be inspected to ensure that excavation is to the correct depth to guard against frost and mechanical damage due to traffic, ploughing or other agricultural activities on open land.

Trenches should not be backfilled until these conditions have been satisfied and the installation is seen to conform to the drawings, specifications and appropriate regulations.

Valve and hydrant boxes should be properly aligned, and suitable valve-operating trays provided before a pipeline is accepted.

Inspection of installations within buildings

All internal pipework must be inspected to ensure that it has been securely fixed.

Before testing takes place all cisterns, tanks, hot water cylinders and water heaters must be inspected to ensure that they are properly supported and secured, that they are clean and free from swarf, and that cisterns are provided with correctly fitting covers.

Water undertakers expect to be notified of certain water installations before the work is carried out, and particularly those with a high contamination risk. Proper notification as in Regulation 5 allows inspectors to discuss situations prior to installation and so avoid problems later. This also enables them to ensure that completed installations comply with relevant requirements.

Visual inspection is an essential part of both interim and final tests and will detect many faults that the formal test will not pick up and which might lead to failure at a later date. Visual inspections should be made before any work is concealed. A careful record should be kept of such inspections and notes should be made to help with the preparation of 'as installed' drawings.

12.2   Testing for soundness

All water supply installations are required by Water Regulations to be tested for soundness. The guidance document is quite specific and states the following:

> 'Installations, including all supply and distributing pipes, fittings and connections to appliances, must be tested hydraulically (water pressure test) for a test period of one hour, at a test pressure of 1.5 times the maximum operating pressure, or the maximum operating pressure plus an allowance for any expected surge, whichever is the greatest. During this, there should be no visible leakage, and no loss of pressure. Normal working pressure for pumping mains should take account of any likely surge pressures.'

BS 6700 gives advice on the testing of installations above and below ground and sets out test procedures for rigid and elastomeric pipes which are given in section 12.3.

Timing of tests

Interim tests should be applied to every pipeline as soon as practicable after completion of that particular section, with particular attention to all work that will be concealed. For buried pipelines these should be carried out before backfilling is placed over joints. However, some backfilling will be necessary to hold the pipes in place and prevent any movement of them during the test period. Long pipelines should be tested in sections as the work proceeds.

Final tests should be carried out on completion of all relevant work. Completion of buried pipelines includes backfilling, compacting and surface finishes. Final tests are generally carried out immediately before the hand-over date, but where the installation area is not likely to be affected by site works, the test may be done as work is completed.

Items failing any test should be corrected immediately and retested before further work proceeds.

12.3 Testing methods

Mains should be pressure tested at 1.5 times the normal working pressure (see figure 12.1). BS 6700 suggests twice working pressures but this may not always be advisable where working pressures are very high or for PVC-U mains, which may retain stresses.

Normal working pressures in pumping mains should take account of any likely surge pressure.

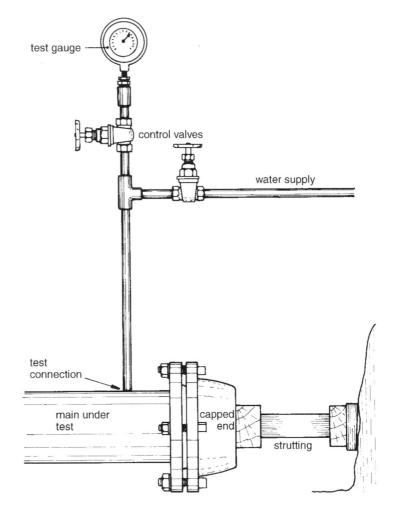

Pressure gauge fixed at lowest point on main.

Check and calibrate pressure gauge before beginning test.

Air valves positioned at high points to release trapped air.

Capped ends strutted securely against solid ground. All bends and branches securely anchored.

Place sufficient backfill before testing to prevent movement of pipes under test.

Allow 30 minutes before test period for water absorption and for water temperatures to warm/cool to ambient temperature.

Bring main up to test pressure slowly and release carefully after testing.

Using mains water fill main slowly to expel air and disconnect from main before beginning test.

Allow time for water absorption and for stabilization of water temperatures before starting test.

Figure 12.1   Testing underground pipelines

Before testing

On above ground installations, all jointing should be completed with pipes and components properly secured before commencement of the test.

Pipes below ground should be fully installed, ends capped and pipes fully anchored to prevent movement under test.

Pipes and components should be inspected for obvious signs of leakage or other irregularities. Valves within the test section should be fully open to ensure the whole section is tested.

Testing procedure for rigid pipe installations

After installation:

(1) Fill the installation slowly with drinking water allowing the air to escape. (It is extremely difficult to pressurize a system containing air pockets.)
(2) Allow to stand for 30 minutes to allow water temperature to stabilize.
(3) Inspect the whole system and its joints visually for leakage.
(4) Pump up to test pressure (as stated above).
(5) Leave to stand for 1 hour.
(6) Check for visible leakage and for loss of pressure. If neither occurs, the test is deemed to be satisfactory.
(7) Otherwise repeat test after locating and repairing any leakage.

Testing elastomeric pipes

Two procedures are shown in BS 6700 for the testing of elastomeric pipes.

Test procedure A (figure 12.2)

(1) Apply test pressure (1.5 times maximum working pressure) by pumping for a period of 30 minutes and inspect visually for leakage.
(2) Reduce pressure by bleeding water from the system to 0.5 times maximum working pressure. Close the bleed valve.
(3) Visually check and monitor for 90 minutes. If the pressure remains at, or above 0.5 times maximum working pressure, the system can be regarded as satisfactory.

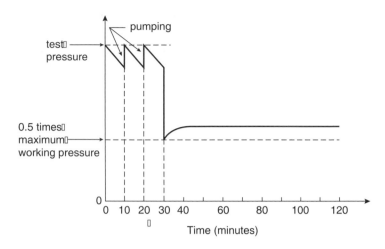

Figure 12.2   **Testing elastomeric pipe systems – test procedure A**

Test procedure B (figure 12.3)

(1)   Apply test pressure (1.5 times maximum working pressure) by pumping for a period of 30 minutes. Note the pressure and inspect visually for leakage.

(2)   Note the pressure after a further 30 minutes. If the pressure drop is less than 60 kPa (0.6 bar) the system can be considered to have no obvious leakage.

(3)   Visually check and monitor for 120 minutes. If the pressure drop is less than 20 kPa (0.2 bar) the system can be regarded as satisfactory.

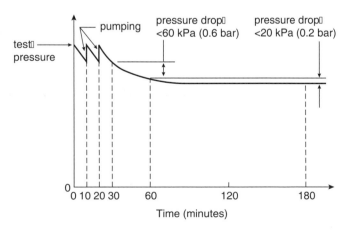

Figure 12.3   **Testing elastomeric pipe systems – test procedure B**

12.4   Flushing and disinfection

It is a requirement of Water Regulations that 'water fittings shall be ...
flushed and where necessary, disinfected prior to use'.

Flushing

All pipes, fittings, appliances, storage vessels and other components
should be thoroughly flushed out with drinking water before being brought
into use. Additionally, if a system is not used immediately before commissioning, it should be flushed at regular intervals and disinfected before
being brought into use.

After flushing, all systems other than single private dwellings should be
disinfected. For single dwellings, thorough flushing is considered to be
sufficient.

Disinfection

Disinfection should always be carried out in the following instances:

- in all new hot and cold installations, appliances and components,
  except single dwellings and other similar small installations;
- in all systems that have been modified or altered in any way;
- on pipes laid below ground, except for minor localized repairs or the
  insertion of tee junctions;
- junctions and fittings used for minor localized repairs should be washed
  or immersed in disinfectant before insertion into the pipeline;
- on any new system that has been standing unused after completion, for
  up to 30 days depending on the characteristics of the water (contact
  water supplier for advice);
- on any system that has been taken out of use or not used regularly;
- on any installation (new or existing) where any form of contamination is
  suspected.

When disinfection processes are carried out on mains pressure installations and there is no backflow device fitted to prevent backflow to the
mains, the water undertaker should be informed. Where water for disinfection is to be discharged to a sewer or water course, the appropriate
authority should be consulted.

BS 6700 shows two disinfection methods:

(1)   using chlorine as the disinfectant; and
(2)   using approved disinfectants other than chlorine. In these cases
      manufacturers' recommendations should be strictly followed.

Any chemicals used should be chosen from those listed in the Drinking
Water Inspectorate's *List of Approved Substances* and published in the
Water Fittings and Materials Directory.

Before commencing the disinfection process, checks should be made to ensure the disinfection liquid will not adversely affect the materials or any protective coatings used in the system.

Safety

It is important that no other chemicals, such as toilet cleansers, are added to any system until cleansing and disinfection is complete as the mixing of chemicals in this way could lead to the generation of toxic fumes.

All users of the building, including temporary and part-time users, e.g. cleaners and security staff, should be notified before the disinfection process is commenced, and notices should be displayed to show that the system and its appliances are out of use. Where possible, affected areas should be closed off.

> **Safety Notice**
>
> # DISINFECTION IN PROGRESS
> # DO NOT USE

The disinfection sequence should be: water mains, supply pipes, cisterns and finally, the distributing system.

For supply pipes, disinfection liquid should be injected through a properly installed injection point.

Procedures for disinfection of both supply pipes and distributing systems are set out in figures 12.4 and 12.5.

Mains and large diameter supply pipes

Piping must be effectively cleansed and disinfected before first being used and after being opened up for repair or alteration (see figure 12.4).

At the time of laying, large pipes should be brushed clean and sprayed internally with a strong solution of sodium hypochlorite. For small diameter pipes insert a polyurethane foam plug soaked in sodium hypochlorite solution at a strength of about 10% chlorine and pass it through the bore.

In pipework under pressure, chlorine should be pumped in through a properly installed injection point at the beginning of the pipeline, until the residual at the end of the pipeline is in excess of 2 ppm (parts per million).

If the pipework is under mains pressure inform the water supplier of the intention to carry out chlorination. All work must be carried out in accordance with the water supplier's requirements.

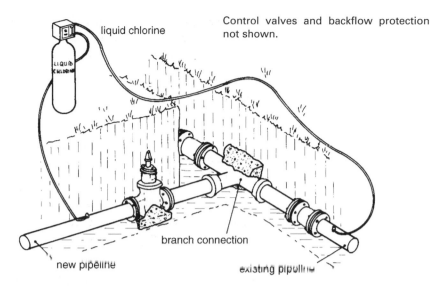

Before connection:
- brush clean and disinfect pipes and fittings before insertion into pipeline;
- disinfect trench around connection before cutting into existing pipeline;
- put chlorine tablet or similar inside pipe before final connection.

Pipe from existing main can be used to fill new main for pressure test and for chlorination purposes, provided precautions are taken against backflow and the water undertaker agrees.

After connection:
- flush out new main using water from existing pipeline;
- fill with chlorinated water (50 ppm);
- allow chlorinated main to stand for 24 hours;
- if free residual chlorine is less than 30 ppm repeat the process; otherwise
- flush out and take samples for chemical and bacteriological analysis. Free residual chlorine to be at level of drinking water supplied.

Figure 12.4   **Disinfection of mains connection**

Disinfecting supply pipe

(1) Thoroughly flush and empty at least twice.

(2) Fill the system with clean water.

(3) Close all draw-off taps, and valves.

(4) Inject chlorine at rate of 50 ppm.

(5) Open draw-offs in turn starting at the bottom and check for residential chlorine at each draw-off point.

(6) Leave for contact period of 1 hour.

(7) After contact period, check residual chlorine at each draw-off point. If less than 30 ppm the disinfection process should be repeated.

(8) Drain and flush system until residual chlorine is at same level as incoming drinking water.

(9) Take sample for bacteriological analysis.

Disinfecting distributing system

(1) Thoroughly flush and empty at least twice.

(2) Fill the system with clean water. Cistern to be full to overflowing level.

(3) Close all draw-off taps, valves and shut servicing valve (A).

(4) Add chlorine to cistern at rate of 50 ppm.

(5) Open draw-offs in turn starting at the top (nearest the cistern) and check for residual chlorine at each draw-off point.
Note: Cistern may need to be topped up as work proceeds.

(6) Leave for contact period of 1 hour.

(7) After contact period, check residual chlorine at each draw-off point. If less than 30 ppm the disinfection process should be repeated.

(8) Drain and flush system until residual chlorine is at same level as incoming drinking water.

(9) Take sample for bacteriological analysis.

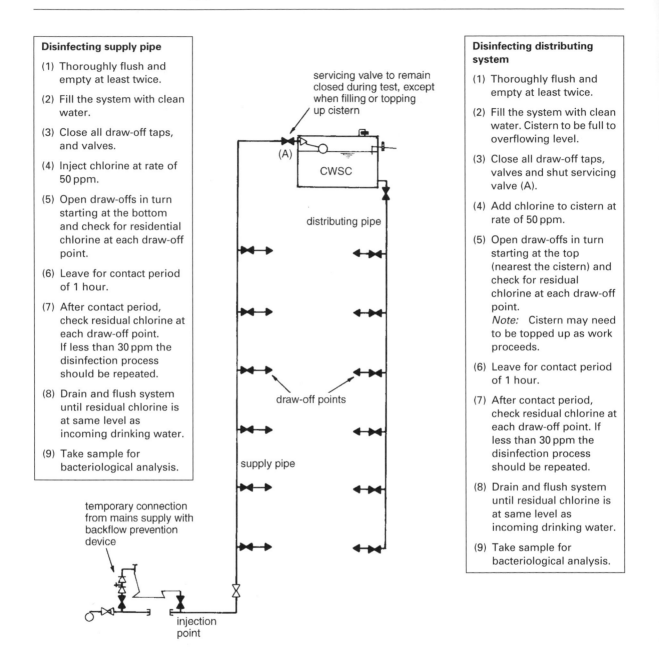

If direct connection from a main is to be made, backflow protection should be provided and the water supplier should be notified.

The water supplier may require an interposed cistern to be used.

Figure 12.5   Disinfection of water services

Installations within buildings

Storage cisterns and distributing pipes must be disinfected using the following method.

Prior to disinfection any users of the building or water supply should be informed and the system, or part of it, put out of use.

The cistern and pipe must first be filled with water and thoroughly flushed out. The cistern must then be filled with water again and a disinfecting chemical containing chlorine added gradually whilst the cistern is filling, to ensure thorough mixing. Sufficient chlorine should be used to give a concentration of at least 50 ppm.

When the cistern is full, the supply should be stopped and all the taps on the distributing pipes must be opened successively, working progressively away from the cistern. Each tap should be closed when the water discharged begins to smell of chlorine. The cistern should then be topped up with water from the supply pipe and with more disinfecting chemical in the recommended proportions. The cistern and pipes should then remain charged for at least 1 hour, after which a test must be made for residual chlorine. If none is found, the disinfecting process must be repeated.

Finally, the cistern and pipes should be thoroughly flushed out with clean water before any water is used for domestic purposes.

If ordinary 'bleaching powder' is used, the proportions must be 150 g of powder to 1000 l of water; the powder should be mixed with water in a separate clean vessel to a creamy consistency before being added to the water in the cistern. For proprietary brands of chemicals, use the proportions recommended by the manufacturer.

For single dwellings and small pipework alterations, flushing is all that is required unless contamination is suspected, in which case the system should be disinfected.

12.5 Maintenance

To keep the performance of any installation at the original specified design standard, it will be necessary to carry out some maintenance work. The amount of maintenance work will depend upon the type and size of system, and the risk or effects of breakdown balanced against the frequency and cost of the inspection and maintenance programme.

Planned preventative maintenance, regularly carried out, will help to ensure that systems perform correctly and avoid most breakdowns and the risk of costly damage to components, equipment and buildings. Table 12.1 shows a typical maintenance schedule.

Building owners should be provided with a maintenance schedule and instructions along with detailed and accurate drawings of the installation including all pipe runs, concealed or otherwise.

Maintenance and repairs should be carried out by a competent person, which means one who has the skills, knowledge and experience of water services installations, and of relevant statutory requirements.

Table 12.1   Maintenance schedule

Component	Maximum time interval	Remarks
Inspections	12 months	At frequent intervals in addition to any statutory inspection.
Water analysis	6 months or as required	Larger buildings (not individual dwellings) particularly where drinking water is stored. Gives useful guide to condition of an installation. Check cisterns, hot and cold water outlets.
Water temperatures	6 months or as required	Check during periods of most severe conditions, e.g. high occupancy, extreme heat or cold.
Cleaning and disinfection	12 months or as required	Following alterations to system and when contamination is suspected.
Meters	6 months	Read meters for consumption and early warning of wastage. Check that meters are working.
Meter and stopvalve chambers	12 months or as required	Inspect to ascertain state of chamber construction. Clean out as necessary, check that box and lid are in working order and undamaged and grease hinges.
Earthing and bonding		Check when other inspections, maintenance or alterations are being made (electrician to check electrical continuity).
Control valves	12 months	Operate to check that they close tightly and operate smoothly. Repair or renew as necessary. Valve keys should be available for emergency use. Valves should be labelled clearly to show what they do.
Pressure relief valves and temperature relief valves	6 to 12 months	Ease at regular intervals to ensure they are not stuck or that outlet is blocked. Remedy faults immediately. Easing valves may sometimes cause them to leak but this is preferable to an explosion. Check discharge pipe is unobstructed.
Pressure reducing valves	6 to 12 months	Check pressures downstream of valve and investigate any changes from normal.
Vessels under pressure	6 months	Inspect water storage vessels and expansion vessels for signs of deterioration. Measure gas pressures and adjust if not within manufacturer's recommended limits.
Pressure gauges	6 months	Investigate any change from normal pressure.
Storage cisterns	6 months	Inspect for cleanliness and clean out as required. Check lid is securely in place. Look for signs of leakage or corrosion. Check linked cisterns for stagnant water (taste, odour or dust on surface). Check condition of bearers, safes and safe outlets. Check overflow pipes for obstruction and correct fall. Check insulation before winter.
Float-operated valves	6 months	Check operation and closing. Adjust for correct water level. Open float valve in feed and expansion cistern to prevent it sticking in the closed position.

continued

Table 12.1   continued

Component	Maximum time interval	Remarks
Terminal fittings	6 to 12 months or as required	Rewasher, reseat or renew taps as required to prevent leakage. Tighten packing glands and check spindles for wear and efficient action. Check self-closing taps at intervals and remedy any faults. Clean sprayheads on taps, showers and shower mixers at intervals depending on rate of furring.
Pipework	12 months	Tighten loose fittings and supports and replace any missing ones. Check provision for expansion and contraction (especially plastics pipework). Check and tighten joints, remake or renew as necessary. Before commencing any work check compatibility of pipes and fittings, i.e.: ○ sizes, and whether imperial or metric; ○ metals for corrosion; ○ plastics for jointing method; ○ materials are available for repairs.
Disused fittings	1 month	Sterilize every 6 months or disconnect at branch connection.
Corrosion	12 months	Inspect for outward signs of corrosion. Reduced flow rates may indicate that corrosion products are causing obstruction. Replace corroded or seriously scaled lead pipe with pipe of some other suitable material. Internal corrosion of galvanized steel pipe is usually localized, although in some waters the zinc coating can break up looking very like sand deposits in cisterns and tanks. In this case replace the complete pipe length. Pitting corrosion of copper pipe may be due to carbon film in pipe bore or cathodic scale. Rapid water velocities may lead to erosion corrosion. Pipes in damp conditions and areas where the air is acidic need regular inspections, and pipes should be protected against corrosive effects. Where pipes enter buildings they are particularly vulnerable at floor level, especially in the back of sink units.
Earth continuity bonding	12 months or as required	Check to be carried out by competent electrician after pipes, fittings or appliances have been removed or replaced, or additions have been made to installation.
Insulation and fire stopping	12 months	Inspect and make good any damage. A good time would be before the start of winter.
Ducting	6 months	Accessibility is essential. Check that ducting is clear of extraneous matter and free from vermin. Check that access is not obstructed and entry is readily possible. Check crawlways and subways for leakage from pipework, entry of ground or surface water, and accumulation of flammable materials.

12.6 Locating leaks

BS 6700 deals with leak detection to premises supplied by meter only. As most premises in this country are unmetered the author has also included notes relating to these.

The first signs of leakage in any premises are usually:

- Visual.
- Noise. The sound of water escaping through a small crack or hole in a pipe is similar to that of a cistern filling, but of course it will be continuous, and will probably be noticed more at night when the pressures are higher and there are fewer extraneous noises.
- In metered supplies the high meter reading or account is often the first indication, which leads to leak detection taking place.

Water authorities periodically carry out their own waste water detection exercises and many leaks are located in this way. Water undertakers have a duty to keep wastage to a minimum, and if asked will usually send an inspector along to advise, and possibly to assist in locating both the service pipe and the leak. However, there are fairly simple ways to establish first that there is a leak, and secondly its approximate location.

Unmetered supplies

Procedure
(1) Look for visible signs of leakage, i.e.:
    - wet or soggy patches on ground,
    - areas of grass that are greener or growing stronger and faster than remainder,
    - water in stopvalve chambers.

(2) Make sure no water is being drawn off or used.

(3) Listen at taps and stopvalves to see where the noise appears loudest. There are excellent pocket stethoscopes available which can be used directly on a tap or on a stopvalve key which in turn is resting on an underground stopvalve. This will often give a general indication of the whereabouts of the leak.

(4) Turn off at the main stopvalve and listen again. If the noise has stopped, it usually means the leak is on the supply pipe. If the noise continues the leak is probably on the communication pipe or main.

    Note It is important that the stopvalve turns off effectively, otherwise the leak may not be where it appears to be.

(5) If there are branch supplies with stopvalves fitted, then the procedure can be repeated by closing each branch supply in turn, to establish if the leak is on the common pipe or one of its branches (see figure 12.6).

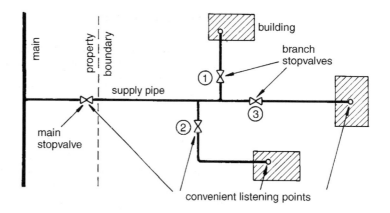

Figure 12.6   **Locating leaks in branch supply pipes**

(6)   It is not always possible to locate the leak exactly, and in long pipelines the best approach may be to cut and plug the pipe about halfway along its length (or fit an intermediate stopvalve) and retest. This can be repeated a number of times to establish which length of pipe needs to be dug up (see figure 12.7).

(7)   It is also possible to drive a pointed steel bar into the ground at intervals along a pipeline, pulling it out again to see how wet it is. Usually it is wetter nearer the leak.

Note: The occurrence of a leak is often an indication of the state of the pipe, especially with steel pipelines.

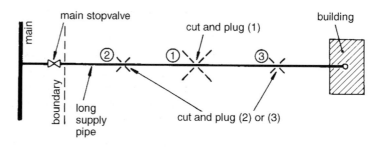

Figure 12.7   **Locating leaks in long pipelines**

Metered supplies

The method shown in BS 6700 follows a simple logical procedure using the meter and a watch, and requires first that the supply pipe and its stopvalves are located and recorded on a diagram. The isolating stopvalves should be numbered. The following example (see figure 12.8), based on that shown in BS 6700, describes the procedure.

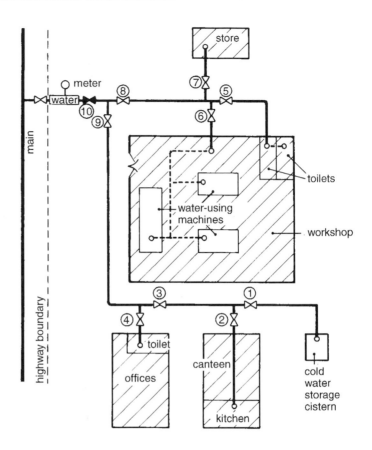

Figure 12.8   **Locating leaks in metered supply – typical pipework network**

Before the test:

- check that all stopvalves are in working order and will stop the supply when closed;
- make sure that no part of the supply is in use and that all water fittings except isolating valves are closed;
- check that the meter is capable of recording low flows.

Procedure
(1) Record rate of flow using the meter and a watch; if zero there is no detectable leakage.
(2) If there is a rate of flow, shut off isolating valves in sequence starting at the valve furthest from the meter, i.e. no. 1 in figure 12.8. After each valve has been shut any change in the rate of flow should be noted.
(3) Leakage should be sought in any sections of the network where closure of the isolating valve has reduced the rate of flow.
(4) As a check, repeat the test procedure when the leakages have been detected and repaired.

Practical example (see figure 12.8)

Item	Rate of flow l/min	Change in rate of flow l/min	Remarks
Start of test	90	—	
Shut valve no. 1	90	Nil	
Shut valve no. 2	45	45	Leakage at 45 l/min (in canteen section)
Shut valve no. 3	45	Nil	
Shut valve no. 4	45	Nil	
Shut valve no. 5	30	15	Leakage at 15 l/min (in toilets section)
Shut valve no. 6	30	Nil	
Shut valve no. 7	Zero	30	Leakage at 30 l/min (in supply to store)
Shut valve no. 8	Zero	Nil	
Shut valve no. 9	Zero	Nil	

12.7   Occupier information

Occupiers and owners of property have an interest in keeping their systems in good order and are in the position, as users, to note the first signs of faults appearing.

Occupiers should be informed of the need for maintenance and be provided with maintenance instructions along with record drawings upon which pipe runs and valves are accurately marked.

British Standards relevant to this book

BS 21	Pipe threads for tubes and fittings where pressure-tight joints are made on the threads (metric dimensions)
BS 417	Specification for galvanized low carbon steel cisterns, cistern lids, tanks and cylinders
	Part 2 Metric units
BS 486	Specification for asbestos-cement pressure pipes and joints (withdrawn)
BS 699	Copper direct cylinders for domestic purposes
BS 750	Specification for undergound fire hydrants and surface box frames and covers
BS 864	Capillary and compression tube fittings of copper and copper alloy
	Part 2 Specification for capillary and compression fittings for copper tubes (obsolescent)
BS 1010	Specification for draw-off taps and stopvalves for water services (screwdown pattern)
	Part 2 Draw-off taps and above ground stopvalves
BS 1192	Construction drawing practice
	Part 1 Recommendations for general principles
	Part 3 Recommendations for symbols and other graphic conventions
BS 1211	Specification for centrifugally cast (spun) iron pressure pipes for water, gas and sewage
BS 1212	Float-operated valves
	Part 1 Specification for piston type float operated valves (copper alloy body) (excluding floats)
	Part 2 Specification for diaphragm type float operated valves (copper alloy body) (excluding floats)
	Part 3 Specification for diaphragm type float operated valves (plastics bodied) for cold water services only (excluding floats)
	Part 4 Specification for compact type float operated valves for WC flushing cisterns (excluding floats)
BS 1387	Specification for screwed and socketed steel tubes and tubulars and plain end steel tubes suitable for welding or for screwing to BS 21 pipe threads
BS 1453	Specification for filler materials for gas welding
BS 1563	Cast iron sectional tanks (rectangular)
BS 1564	Pressed steel sectional rectangular tanks
BS 1565	Specification for galvanized mild steel indirect cylinders, annular or saddle-back type
	Part 2 Metric units
BS 1566	Copper indirect cylinders for domestic purposes
	Part 1 Specification for double feed indirect cylinders
	Part 2 Specification for single feed indirect cylinders
BS 1710	Specification for identification of pipelines and services
BS 1845	Specification for filler metals for brazing
BS 1968	Specification for floats for ballvalves (copper)
BS 1972	Specification for polythene pipe (Type 32) for above ground use for cold water services (withdrawn)
BS 2035	Specification for cast iron flanged pipes and flanged fittings
BS 2456	Specification for floats (plastics) for float operated valves for cold water services
BS 2494	Specification for elastomeric seals for joints in pipework and pipelines
BS 2580	Specification for underground plug cocks for cold water services

BS 2779	Specification for pipe threads for tubes and fittings where pressure-tight joints are not made on the threads (metric dimensions)
BS 2782	Methods of testing plastics
	Part II Thermal plastics pipes, fittings and valves
BS 2871	Specification for copper and copper alloys. Tubes
	Part 1 Copper tubes for water, gas and sanitation (withdrawn)
BS 2879	Specification for draining taps (screw-down pattern)
BS 3251	Specification for indicator plates for fire hydrants and emergency water supplies
BS 3284	Specification for polythene pipe (Type 50) for cold water services (withdrawn)
BS 3456	Specification for safety of household and similar appliances
BS 3505	Specification for unplasticized polyvinyl chloride (PVC-U) pressure pipes for cold potable water
BS 3506	Specification for unplasticized PVC pipe for industrial uses
BS 3955	Specification for electrical controls for household and similar general purposes
BS 4127	Specification for light gauge stainless steel tubes, primarily for water applications
	Part 2 Metric units
BS 4213	Specification for cold water storage and combined feed and expansion cisterns (polyolefin or olefin copolymer) up to 500 l capacity used for domestic purposes
BS 4346	Joints and fittings for use with unplasticized PVC pressure pipes
BS 4346	Part 1 Injection moulded unplasticized PVC fittings for solvent welding for use with pressure pipes, including potable water supply
	Part 2 Mechanical joints and fittings, principally of unplasticized PVC
	Part 3 Specification for solvent cement
BS 4622	Specification for grey iron pipes and fittings
BS 4772	Specification for ductile iron pipes and fittings (withdrawn)
BS 4991	Specification for propylene copolymer pressure pipe
BS 5154	Specification for copper alloy globe, globe stop and check, check and gate valves
BS 5163	Specification for predominately key-operated cast iron gate valves for waterworks purposes
BS 5292	Specification for jointing materials and compounds for installations using water, low-pressure steam or 1st, 2nd and 3rd family gases (obsolescent)
BS 5306	Fire extinguishing installations and equipment on premises
BS 5412	Specification for the performance of draw-off taps with metal bodies for water services
BS 5422	Method of specifying thermal insulating materials on pipes, ductwork and equipment (in the temperature range $-40°C$ to $+70°C$)
BS 5433	Specification for underground stopvalves for water services
BS 5556	(withdrawn – replaced by BS 2782 – Part 2)
BS 5728	Measurement of flow of cold potable water in closed conduits
BS 6076	Tubular polymeric film for use as protective sleeving for buried iron pipes and fittings (for site and factory application)
BS 6091	Vulcanized fibre for electrical purposes
	Part 1 Specification for general purposes
	Part 2 Methods of test
BS 6144	Specification for expansion vessels using an internal diaphragm, for unvented hot water supply systems
BS 6281	Devices without moving parts for the prevention of contamination of water by backflow
	Part 1 Specification for type A air gaps
	Part 2 Specification for type B air gaps
	Part 3 Pipe interrupters of nominal size up to and including DN 42
BS 6282	Devices with moving parts for the prevention of contamination of water by backflow
	Part 1 Specification for check valves of nominal size up to and including DN 54
	Part 2 Specification for terminal anti-vacuum valves of nominal size up to and including DN 54
	Part 3 Specification for in-line anti-vacuum valves of nominal size up to and including DN 42

	Part 4 Specification for combined check and anti-vacuum valves of nominal size up to and including DN 42
BS 6283	Safety devices for use in hot water systems
	Part 1 Specification for expansion valves for pressures up to and including 10 bar
	Part 2 Specification for temperature relief valves for pressures from 1 bar to 10 bar
	Part 3 Specification for combined temperature and pressure relief valves for pressures from 1 bar to 10 bar
	Part 4 Specification for drop-tight pressure reducing valves of nominal size up to and including DN 54 for supply pressures up to and including 12 bar
BS 6351	Electric surface heating
	Part 1 Specification for electric surface heating devices
BS 6437	Specification for polyethylene pipes (type 50) in metric diameters for general purposes
BS 6465	Sanitary installations
	Part 1 Code of Practice for scale of provision, selection and installation of sanitary applicances
BS 6572	Specification for blue polyethylene pipes up to nominal size 63 for below ground use for potable water
BS 6675	Specification for servicing valves (copper alloy) for water services
BS 6700	Specification for design, installation, testing and maintenance of services supplying water for domestic use within buildings and their curtilages
BS 6730	Specification for black polyethylene pipes up to nominal size 63 for above ground use for cold potable water
BS 7206	Specification for unvented hot water storage units and packages
BS 7291	Thermoplastics pipes and associated fittings for hot and cold water for domestic purposes and heating installations in buildings
	Part 1 General requirements
	Part 2 Specification for polybutylene (PB) pipes and associated fittings
	Part 3 Specification for cross-linked polyethylene (PE-X) pipes and associated fittings
	Part 4 Specification for chlorinated polyvinyl chloride (PVC-C) pipes and associated fittings and solvent cement
BS 7671	Requirements for Electrical Installations (IEE Wiring Regulations)
BS EN 545	Ductile iron pipes, fittings, accessories and their joints for water pipelines. Requirements and test methods
BS EN 598	Ductile iron pipes, fittings, accessories and their joints for sewerage applications. Requirements and test methods
BS EN 969	Specification for gas (ductile iron pipes)
BS EN 1057	Copper and copper alloys, seamless, round copper tubes for water and gas in sanitary and heating appliances
BS EN 60335	Specification for safety of household and similar electrical appliances
BS EN 60335-2-35	Instantaneous water heaters
	Part 101 General requirements
	Part 102: Section 102.21 Storage water heaters
BS EN 60335-2-51	Stationary circulation pumps for heating and service water requirements
CP 312	Plastics pipework (thermoplastics materials)
	Part 1 General principles and choice of material
	Part 2 Unplasticized PVC pipework for the conveyance of liquids under pressure
	Part 3 Polyethylene pipes for the conveyance of liquids under pressure

References

Backsiphonage Report, HMSO, (1974)

Building Regulations, HMSO

Design guide, Institute of Plumbing (out of print)

Guidance on the application and interpretation of the Model Water byelaws, Department of the Environment, (1986)

Legionaire's Disease – Good Practice Guide for Plumbers, The Institute of Plumbing, (1990)

Model Water bylaws, HMSO, (1986)

N.II IG Publication No. 6. Service entries for new dwellings on residential estates.

Plumbing engineering services design guide, Institute of Plumbing

Principles of laying water mains, Water Authorities Association

Service cores in high flats – cold water services, Ministry of Housing and Local Government, (1965)

Water fittings and materials directory, Water Regulations Advisory Scheme

Water Regulations Guide, Water Regulations Advisory Scheme (WRAS)

Water Supply (Water Fittings) Regulations 1999, 'Regulator's specification for the performance of WC suites', HMSO, (1999)

Water Supply (Water Fittings) Regulations 1999, 'Regulator's Specification on the Prevention of Backflow', HMSO, (1999)

Water supply byelaws guide, Water Research Centre in association with Ellis Horwood Ltd, Chichester

Water Supply (Water Fittings) Regulations 1999. HMSO

Water Industry Act 1991, HMSO

Health and Safety Executive (1992) *'Safe' hot water and surface temperatures.* HS (G) 104. ISBN 0-11-321404-9.

Acts   Water Act 1945
etc.   Water Supply (Water Fittings) Regulations 1999
        Water Industry Act 1991
        Water (Scotland) Act 1980
        Water and Sewerage Services (Northern Ireland) Order 1973
        Building Regulations
        Building Act 1984
        Building Standards (Scotland) Regulations
        Building (Northern Ireland) Regulations
        The Health and Safety at Work etc. Act 1974
        The Workplace (Health, Welfare and Safety) Regulations
        The Gas Safety (Installation and Use) Regulations
        Asbestos Regulations 1969

Index